DATA-STAT

RESPIRATORY

THERAPY

DESK REFERENCE

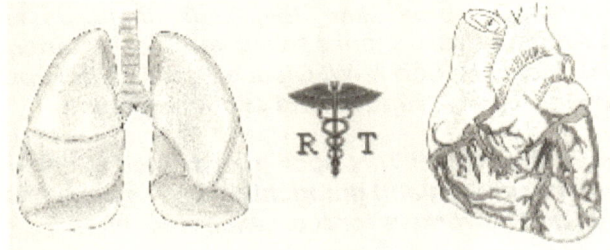

By: Helen Schaar-Corning, RRT
& Stanley Louis Bryant Jr., CRT

The authors of Mosby's Respiratory PDQ,
& numerous other Medical/Health publications

A Fast, Precise, & Comprehensive Guide for Respiratory
Care Professionals. Great for fast facts & exam review.
Contains all the clinically relevant & vital information
found in much larger RT books.
This is the ultimate STAT reference manual
for the desk or locker of RRT's & CRT's.

ISBN # 978 - 1 - 4303 - 1095 - 2

Acknowledgements:

Helen Schaar-Corning, RRT: This book is dedicated with love, to my family (Mom, Grandma, Mike, Joe, Paula, Jamie, Jasen, Michael, & Anna). Special thanks to my very dearest friends: Wayne (W3), Vickie M, Lori M-W, Glenda B, & Kim D. I appreciate your kindness, loyalty, support and encouragement!

Stanley Louis Bryant, Jr.: This book is dedicated to my loving wife Wendy and our children and grandchildren. A special thanks to my patients, and coworkers for their encouragement.

DATA-STAT RESPIRATORY THERAPY DESK REFERENCE

INTRODUCTION

The field of pulmonary medicine continues to grow rapidly, encompassing an ever-expanding scope of care. Complete memorization of all this data would be very difficult, to say the least. Every professional clinician needs all the vital information at their fingertips instantaneously. This book is a comprehensive and precise manual of all the important information vital to safe and effective care of the adult pulmonary patient. Neonatal/Pediatric Respiratory Care is also concisely outlined. This book is specially designed to facilitate finding information with speed and accuracy.

All the newest medical information, and state-of-the-art medical technology is included. The data is organized in fast access charts and tables, plus encyclopedic references. The comprehensive charts contain the vital facts and formulas that are most frequently utilized.

Chapters include diseases, medications, diagnostic tests, PFT's, X-ray's, bronchoscopy, therapeutic modalities, laboratory values, hemodynamics, mechanical ventilation, ACLS, and much more.

The DATA-STAT RESPIRATORY THERAPY DESK REFERENCE is the ultimate STAT reference manual for the desk or locker of RRT's & CRT's. It's fast, precise, and comprehensive; great for fast facts & exam review. This book contains all the clinically relevant & vital information found in much larger RT books.

Please note this book is available at Lulu.com in your choice of full-color or black & white. A PDF download version is also available. The price at Lulu.com is almost always lower than other booksellers. All retailers of this book offer a top quality product.

TABLE OF CONTENTS

9. Laboratory Tests: Blood, Urine, CSF, Sputum

10. Bronchoscopy

11. Chest Tubes

12. EKG's

13. PFT's

14. X-rays

15. Artificial Airways & Suctioning
    Intubation & Extubation Guidelines
    Oral & Nasal Airways
    ET Tubes w/ Tube & Blade Sizes, Cuff Pressures
      Types of Cuffs & Tubes
    EOA, PMV
    Suctioning Guidelines, Vacuum pressures

16. Mechanical Ventilation, CPAP & BIPAP

17. Neonatal/Pediatric Respiratory Care

18. BLS & ACLS

19. Equipment Disinfection

20. Misc. Formulas & Conversions
    Gas Laws
    Temperature Conversions
    Torr, cmH$_2$0, mmHg, & kPaConversions
    Metric Conversions
    Height & Weight, IBW
    BSA, BMI, BMR, REE

ABBREVIATION KEY

B2	Beta2 (as in specificity for Beta2 receptors which cause bronchodilation)
BBS	bilateral breath sounds
BID	two times daily
BSA	body surface area
cap	capsule
CHF	congestive heart failure
cm	centimeters
CNS	central nervous system
CO	Cardiac output
cont	continuous
CVA	cerebrovascular accident or stroke (blood clot or hemorrhage brain blood vessel)
dL	deciliter
DPI	dry powder inhaler
exp	expiratory
f	frequency or rate
f/b	followed by
g	gram
GI	gastrointestinal
hr	hour
ICH	intracranial hemorrhage
IM	intramuscular
in or "	inch(es)
insp	inspiratory
IV	intravenous
kg	kilogram
L	liter
L/m	Liters/minute
L/sec	Liters/second
lb	pound
m^2	meters squared
maint	maintenance
max	maximum
mcg	microgram
MDI	metered dose inhaler

med	medication
mEq	milliequivalent
mfg	manufacturer
mg	milligram
MI	Myocardial infarction
min	minute
mL	milliliter
mm	millimeter
mm Hg	millimeters of mercury or Torr
mol	mole
neb	nebulizer
NS	normal saline 2.5 to 3 mL
NSAID	nonsteroidal anti-inflammatory drug
oz	ounce or ounces
PB	pressure barometric
PO	by mouth
PRN	as needed
psi	pounds per square inch
psr	pressure
pt	patient
QD	once every day
QHS	nightly at hour of sleep
QID	four times daily
resp	respiratory
SQ	subcutaneous
sol	solution
spont	spontaneous
Tab	tablet
TID	three times daily
tx	treatment
μ	micro
μg	microgram
U	units
UTI	urinary tract infection
vent	ventilator
vol	volume
w/	with
w/a	while awake

w/o	without
WOB	work of breathing
Δ	change
↑	increased
↓	decreased

NOTES:

_____

_____

_____

_____

_____

_____

_____

Chapter 1

PULMONARY ANATOMY & ASSESSMENT OF THE PULMONARY PT.

Respiratory Anatomy Topography & Vocal Cords

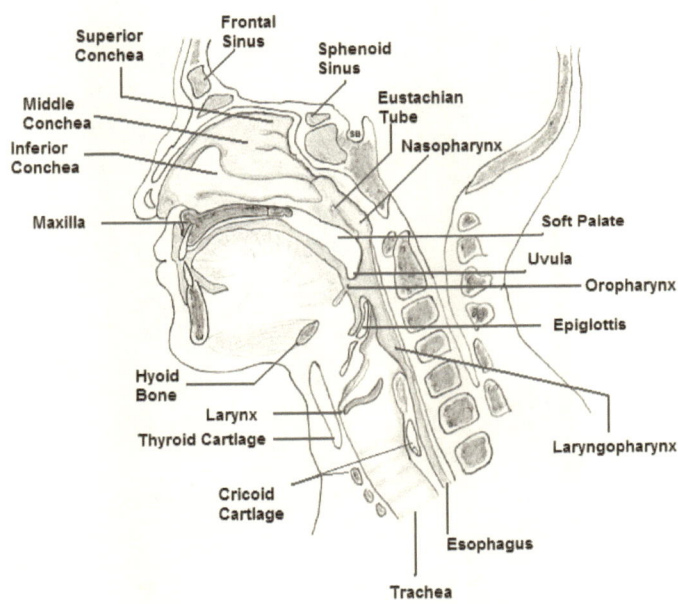

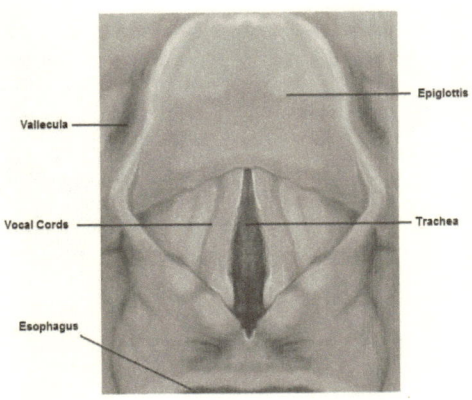

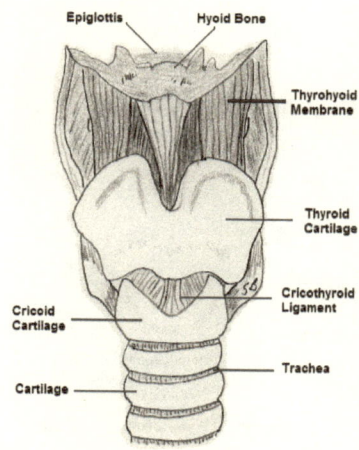

Epiglottis   Hyoid Bone

Thyrohyoid
Membrane

Thyroid
Cartilage

Cricothyroid
Ligament

Cricoid
Cartilage

Trachea

Cartilage

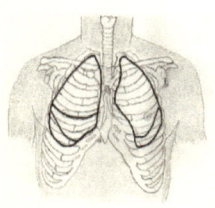

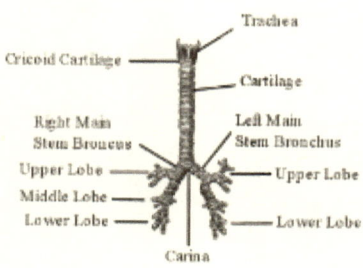

Trachea

Cricoid Cartilage

Cartilage

Right Main
Stem Bronchus

Left Main
Stem Bronchus

Upper Lobe

Upper Lobe

Middle Lobe

Lower Lobe

Lower Lobe

Carina

10

The Lungs

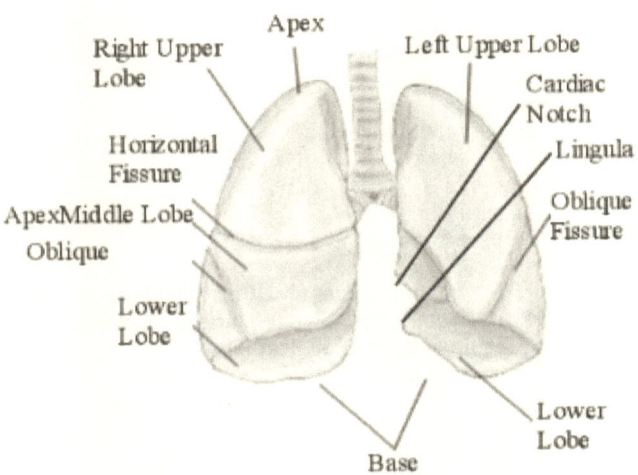

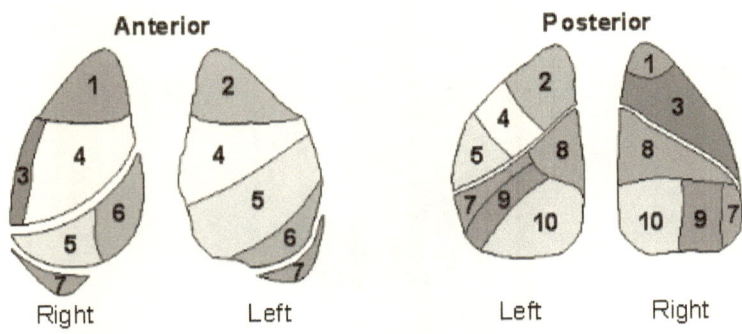

Anterior

Right   Left

Posterior

Left   Right

Lung Segments:

Right Lung

Superior Lobe	1 Apical
	3 Posterior
	4 Anterior
Middle Lobe	5 Lateral
	6 Medial
Inferior Lobe	7 Anterior Basal
	8 Superior

9 Lateral Basal
10 Posterior Basal

Left Lung

Superior Lobe 2 Apical Posterior
4 Anterior
5 Superior Lingula
6 Inferior Lingula

Inferior Lobe 8 Superior
9 Lateral Basal
10 Posterior Basal

The right side of the lung is divided by the oblique and horizontal fissures into three lobes; the upper, middle and lower lobes. The left lung is divided by the oblique fissure into only two lobes; the upper and lower lobes.

The Heart

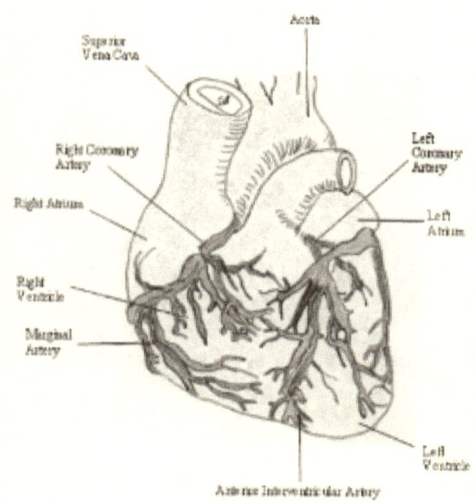

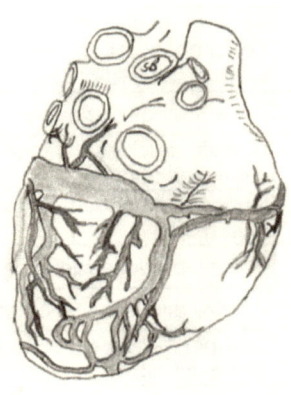

Vital Signs: HR, BP, RR, Temp

Vital Sign	Normal Value	Importance of Values
Heart Rate (HR)	60-100 bpm	Bradycardia < 60 Tachycardia > 100
Blood Pressure (BP)	120/80 Systolic 90-130 Diastolic 60-90	Increased =Hypertension Decreased=Hypotension
Respiratory Rate (RR) (See Resp Pattern below)	8-20/minute	Bradypnea < 8/min Tachypnea > 20/min
Temperature*	Axillary 97.6 °F.......36.4 °C Oral 98.6 °F.......37.0 °C Tympanic 99.6 °F.......37.6 °C Rectal 99.6 °F.......37.6 °C	Hypothermia= low temp Hyperthermia= high temp

Temperature + or − 1 °F is considered normal.
*Also see Hemodynamics & ABG chapters.

Auscultation & Other Assessments

Auscultation & Assessments	Information
Breath sounds	Clear over all lobes is normal.
Bronchophony	Increased tissue density, consolidation
Crackles	Fine crackles (also called rales) indicate fluid in the lungs, pulmonary edema/CHF. Coarse crackles associated w/secretions.

Diminished breath sounds	Poor air movement, fluid in lung tissue, hemothorax, pneumothorax, COPD, or other disease/disorder.
Egophony	e-e-e sounds like a-a-a
Pleural friction rub	Audible crackles at pleural area, that do not change with cough.
Tracheal Sounds	Upper airway edema, stridor, restriction, obstruction.
Vocal fremitus	Voice vibrations radiating into lungs.
Wheeze & Rhonchi	Narrow airways & secretions.
Whispering pectoriloquy	Whisper 1-2-3 with consolidation sounds high-pitched.
Accessory Muscles	Use of accessory muscles including neck muscles, chest & abdominal muscles indicates dyspnea; possibly asthma or COPD exacerbation.
Percussion of chest	Low pitched and dull sounding is normal.

Glasgow Coma Scale

Observation Test		Eyes (Opening)	Verbal Response	Motor Response
	6			Obeys verbal orders
	5		Oriented	Localizes pain
	4	Spontaneous	Confused	Flexion withdrawal to pain
Score	3	To speech	Inappropriate	Flexion abnormal
	2	To pain	Incomprehensible	Extension to pain
	1	None	None	None

Total Score <8 intubate

Breathing Retraining Exercises

With severe COPD, most patients find they take small, quick, shallow breaths. Focusing on breathing and practicing diaphragmatic and pursed-lip breathing will assist the patient in taking deeper breaths. This can help to reduce dyspnea, reduce the severity and frequency of dyspneic episodes, & improve exercise tolerance.

Diaphragmatic breathing helps the lungs expand so that they take in more air. The patient should lie as supine as possible & with one hand on the abdomen & the other hand on the chest. While taking a breath in, press on the abdomen and extend the abdomen outward as far as possible during inhalations. The hand over the abdomen should move outward while the hand over the chest should not move. Once the patient has mastered the procedure while supine, they should practice doing it while sitting or standing. A good exercise program should last 20 – 30 minutes BID or TID.

Pursed-lip breathing helps the patient move air out slower & breathe deeper. The patient should breathe in through the nose (4 – 5 seconds) and out through the mouth (6 seconds) while pressing their lips together. Pursed-lip breathing decreases dyspnea and improves the ability to exercise.

Respiration Patterns

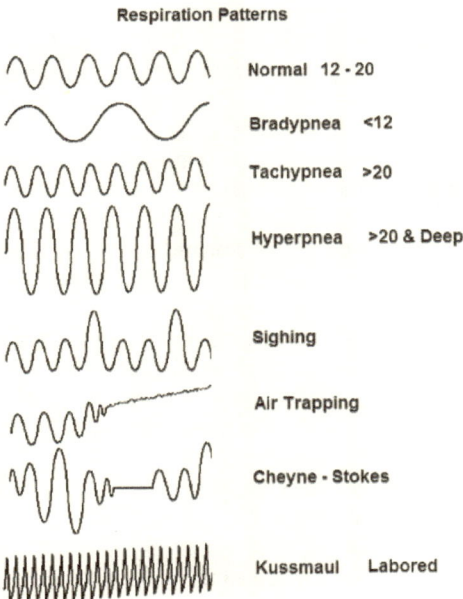

Normal 12 - 20

Bradypnea <12

Tachypnea >20

Hyperpnea >20 & Deep

Sighing

Air Trapping

Cheyne - Stokes

Kussmaul Labored

Chapter 2

PULMONARY DISEASES

ALSO SEE: *Microbes: Bacterial, Viral, & Fungal – Chapter 3*
Neonatal/Pediatric Pulmonary Diseases – Chapter 17

NOTATIONS:
S/S: **Signs & Symptoms**
Tests: **Diagnostic Tests** (emphasizing respiratory tests)
Tx: **Treatment** (highlighting respiratory treatments)

AIDS (Acquired Immune Deficiency Syndrome)

AIDS is defined as infection w/ Human Immunodeficiency Virus (HIV; a retrovirus), w/concurrent AIDS-defining illness. HIV infects T-helper (CD4$^+$) lymphocytes & other body cells resulting in profound immunosuppression & opportunistic infections. This includes: PCP (Pneumocystis carinii pneumonia), tuberculosis, Kaposi's sarcoma, Non-Hodgkins lymphoma B-Cell, coccidiomycosis, histoplasmosis, & many more. AIDS is acquired & spread by contact w/ blood & body fluids as in sexual contact, sharing needles, blood transfusions, occupational exposure, & mother to baby.

S/S: Fever, swollen glands, severe weakness, persistent diarrhea or bloody stools, unexplained bleeding, prolonged loss of appetite, weight loss, leukopenia, & frequent opportunistic infections. Signs can take years to develop. AIDS often develops within 10 years, but can take longer.

Tests: Within 6 months, detectable antibodies develop. Antibody tests are the ELISA & Western Blot. The CD4$^+$ cell count is also diagnostic. Normal CD4$^+$ is 1000 cells/cm^3. ↓ CD4$^+$ indicates ↓ immunity. When CD4$^+$ is < 200 cells/cm^3 opportunistic infections develop. Viral load will ↑ above normal of zero, indicating replication of HIV in the blood.

Tx: Prophylaxis. Prompt treatment of primary & secondary infections & other complications. Pulmonary hygiene, bronchodilators, oxygen, & mechanical ventilation if indicated. Antibiotics, antiretroviral therapy, protease inhibitors, non-nucleoside reverse transcriptase inhibitors (NNRTIs), nucleoside analog reverse transcriptase inhibitors (NRTIs). Drugs include: pentamidine, Zidovudine (AZT), delavirdine, nevirapine, efavirenz, ritonavir, saquinivir, indinavir, & nelfinavir. Cancers are treated w/ chemotherapy, radiation, alpha interferon, surgical resection, lobectomy or pneumonectomy.

Adenoma – See Bronchogenic Carcinoma

Adult Respiratory Distress Syndrome (ARDS)
Restrictive pulmonary disease w/ nonspecific lung response to stress or injury of pulmonary or non-pulmonary nature. Also known as noncardiogenic pulmonary edema, shock lung, white lung & ventilator lung. The A-C membrane is damaged, which results in: insufficient surfactant activity, ↓ compliance, atelectasis, shunt, severe hypoxemia, pulmonary edema, pulmonary fibrosis, ↓ FRC, & respiratory failure. Many causative factors including: oxygen toxicity, prolonged mechanical ventilation, aspiration pneumonia, acute lung injury (ALI), sepsis, viremia, thoracic trauma, hemorrhagic, shock, pancreatitis, air or fat emboli, & near-drowning.
S/S: Respiratory distress, dyspnea, tachypnea, tachycardia, cyanosis, hypoxemia. Breath sounds: wheezes & rales.
Tests: PFT: restrictive disease, decreased FRC. X-ray: ground glass/ honeycomb appearance, diffuse infiltrates, interstitial pattern.
Tx: Treat underlying cause & complications. Mechanical ventilation as required to treat hypercarbia & acidosis; utilize PEEP/CPAP, & lowest acceptable pressures to minimize barotrauma. Oxygen at lowest acceptable level to avoid O_2 toxicity while maintaining adequate PaO_2. Steroids, antibiotics, diuretics, ECMO, surfactant therapy (experimental).

Alpha₁ Antritypsin Deficiency – See Emphysema

Alveolar Proteinosis
Idiopathic, progressive, chronic deposition of lipoprotein into alveoli.
S/S: Ranges from asymptomatic, to mild illness, to fatal. Dyspnea, cough, digital clubbing, & hypoxemia.
Tests: X-ray: diffuse bilateral infiltrates.
Tx: Bronchial hygiene w/ CPT or vibratory PEP therapy & effective cough technique. Bronchoscopy w/ bronchoalveolar lavage. Treat symptoms & complications.

Amyotrophic Lateral Sclerosis (ALS) or Lou Gehrig's Disease
Degenerative neuromuscular disease. Progressive degeneration of the brain & spinal cord motor neurons. Progressive atrophy of the hands, forearms, legs, & eventually withering of muscles throughout the body, including the muscles of ventilation. Poor prognosis w/ 80% of patients dying within 5 years, 5% can live 20 years. Pulmonary complications frequently the cause of death. 8-10% of cases are familial, the others are sporadic. Disease occurrence rate ↑ w/ age.
S/S: Muscle atrophy, muscle weakness, decreased muscle strength, hypoventilation, ineffective cough, atelectasis, pneumonia, hypoxemia, hypercapnia, & respiratory failure.

Tests: Comprehensive testing includes: EMG, NCV (nerve conduction velocity), spinal tap, neuro exam, labs, PFT- restrictive disease w/ ↓ FVC, ↓ PImax, ↓ PEmax.
Tx: Prevent respiratory complications & provide ventilatory support via BIPAP or full support as required. Oxygen. Secretion clearance, cough techniques, & assisted cough.

Aspergillosis

Disease resulting from pulmonary infection w/ Aspergillus fumigatus, a fungal mold found in decaying vegetation, grain & soil. Can be inhaled or ingested. Immunosuppressed patients at higher risk.
S/S: Symptoms common to all types of Aspergillus infection are cough & recurrent pulmonary infections. Allergic aspergillosis is the most common type w/ asthma-type symptoms, eosinophilia, abnormal bronchograms, & ↑ serum IgE level. 2nd disease form is invasive granuloma of the lungs w/ necrotizing pneumonia-type symptoms & hemoptysis. 3rd disease form is infection by fungal ball, mycetoma or aspergilloma, in lung cavities already infected w/ tuberculosis.
Tx: Depends on type & severity. Antifungal medication. Bronchial hygiene therapy, infection control. Lung resection if indicated.

Asthma

An obstructive pulmonary disease characterized by: bronchospasm, reversible airway obstruction, hyper-reactive airways, inflammation, narrowed airways, mucosal edema, ↑ mucus secretions & mucus plugs. Episodes are intermittent & can be mild, moderate, or severe. Various stimuli cause hyper airway responsiveness. Status asthmaticus is severe, life-threatening airway obstruction due to acute asthma episode that is unresponsive to optimal treatment over time.
The etiology is often hereditary. The child of an asthmatic parent is 3 - 6 times more likely to develop asthma. However, anyone can develop asthma at any time in their lives.
Intrinsic – (Nonallergic Asthma): Emotional states of stress or anxiety. Environmental factors like cold, change in temperature or change in humidity. Also, exercise, infection & aspirin.
Extrinsic – (Allergic Asthma): ↑ sensitivity to allergens & irritants. Smoke, air pollution, pollen, certain foods, certain animals, dust mites, infection, & changes in temperature or humidity. Antigen-antibody reaction occurs involving interaction w/ immunoglobulin E (IgE), resulting in degranulation of mast cells. Ruptured mast cells release chemical mediators like histamine, kinins, serotonin, & slow-reacting substance of anaphylaxis (SRS-A). Mediators have potent inflammatory effects causing mucosal edema, bronchospasm, ↑ mucus production & mucus plugs.
S/S: Often brought on by specific triggers or allergens. Intermittent episodes of cough, dyspnea & tachypnea. Also chest congestion, tightness & pain. Also, restlessness, anxiety, orthopnea, use of

accessory muscles, hypoxemia & cyanosis. Breath sounds w/ wheezes, rhonchi. Hypocapnea usually present early in asthmatic episode. Eucapnea in middle stage of episode can indicate impending respiratory failure. Profound ↑ WOB can eventually lead to fatigue, hypoventilation, hypercapnia, & respiratory failure.

Tests: Labs- WBC ↑ if infection present. Eosinophil count ↑ in extrinsic asthma. PFT- obstructive disease & decreased flow rates. FVC & peak flow significantly decreased, & may improve post bronchodilator.

Tx: Prevention & prophylactic treatment. Avoid known allergens & triggers. Pulmonary education & rehabilitation covering the scope of the disease: prevention, therapeutics & medications. Fast-acting beta 2 bronchodilator as needed for symptoms or for prevention prior to exercise, or prior to exposure to known allergen or trigger. Other medications include: Long-acting beta 2 agonist, ipratropium bromide, corticosteroids, cromolyn, nedocromil, theophylline, & leukotriene modifier.

Atelectasis

Poor lung aeration, poor lung expansion, &/or collapse of portion(s) of the lung. Many causes include: lung diseases, severely diminished ventilation, pneumothorax, pleural effusion, tumor, decreased surfactant activity, bronchial obstruction, foreign body, chest trauma.

S/S: Dyspnea, tachypnea, chest pain, tachycardia. Breath sounds- decreased over affected area. Percussion- dull over affected area.

Tests: X-ray reveals location & severity w/ radiopacity, platelike or patchy infiltrates, displaced fissures, possible mediastinal shift toward affected side, possible elevated diaphragm on affected side. PFT- restrictive pattern.

Tx: Treat underlying cause. Improve aeration. Bronchial hygiene.

Blastomycosis

Disease caused by Blastomyces dermatitides, a yeastlike fungal organism. Primarily limited to North America. Found in soil. Infection acquired by inhalation. Results in granulomatous systemic infection of the skin, bones, lungs & other organs. Can produce systemic infection, or respiratory infection w/ lesions, lobar pneumonia, pleural effusions, miliary process & lymph node enlargement.

S/S: Can be an asymptomatic self-limiting illness. Others experience fever, chest cold, cough w/ hemoptysis or purulent sputum, dyspnea, chest pain, & headache.

Tests: X-ray: possible lobar consolidation w/ cavitation.

Tx: Antifungal medication. Bronchial hygiene therapy.

Bronchiectasis

Obstructive pulmonary disease w/ destructive changes in bronchial wall, loss of resiliency, & abnormal, irreversible dilation of bronchi & bronchioles. Leads to chronic inflammation, fibrosis, necrosis, &

19

impaired mucus clearance. There are 3 types of bronchiectasis: Saccular (cystic), Cylindrical (fusiform), & Varicose. Saccular is the most severe. Bronchopulmonary infection in early childhood is the most common cause. Also: Chemical irritants, bacterial infection, pneumonia, bronchial asthma, tumors, atelectasis, cystic fibrosis, carcinoma, & immune deficiency diseases.

S/S: Breath sounds- wheezes, rhonchi, rales. Chronic loose productive cough w/ copious amount of mucopurulent sputum that may separate into 3 layers. Top layer- frothy & watery; center layer-turbid & mucopurulent; bottom layer- opaque & purulent. Also, hemoptysis, frequent pulmonary infections, cyanosis & weakness.

Tests: X-ray reveals type of bronchiectasis: saccular- widespread atelectasis; cylindrical- hazy vascular markings. Hyperinflation of unaffected areas. Bronchogram reveals location & severity w/ bronchi clearly outlined (air bronchograms). ABG- hypoxemia & hypercapnea in advanced disease. PFT- obstructive disease pattern.

Tx: Bronchial hygiene, CPT, PD, aerosol therapy, mucolytics, & bronchoscopy. Antibiotics for infection. Avoid smoke & other irritants. Maintain good nutrition. If other methods fail, surgical resection for severe localized bronchiectasis.

BOOP (Bronchiolitis Obliterans w/ Organizing Pneumonia) – see Pneumonia

Bronchitis

Inflammation of bronchial mucosa that can be acute or chronic. Chronic bronchitis (blue bloater) is classified as COPD. Chronic bronchitis is defined as: 3 months or longer of cough w/ mucus production most days, in episodes that recur for 2 successive years. Bronchitis can be hereditary or brought on by environmental factors. Acute bronchitis associated w/ viral or bacterial infection or inhalation of chemicals, pollutants, or irritants. Chronic bronchitis associated w/ long-term exposure to inhaled smoke, other pollutants, occupational exposure to chemicals, chronic asthma, & frequent pulmonary infections. Greatly ↑ airway resistance, w/ air-trapping.

S/S: Acute bronchitis – Cough, initially nonproductive, later productive. Breath sounds: rhonchi, wheezes. Can lead to bronchiolitis, pneumonia, & chronic bronchitis.

Chronic bronchitis - Cough w/ copious amount of mucopurulent sputum, dyspnea, cyanosis. ↑ use of accessory muscles, pursed-lip breathing, & digital clubbing possible. Cor pulmonale can develop. Breath sounds: rhonchi, wheezes.

Tests: PFT: obstructive disease pattern. ABG: possible hypoxemia, hypercapnia, compensated respiratory acidosis, & acute or chronic respiratory failure. X-ray: hyperinflation & depressed diaphragm.

Tx: Avoid smoke & other pollutants. Expectorants, cough suppressants, humidification of inspired air, oxygen, bronchial

20

hygiene, bronchodilators. Anti-inflammatory medication, steroids.
Antibiotics for infection. Additional therapy for chronic bronchitis
includes: low-concentration oxygen (with consideration for hypoxic
drive breathing). Diuretics & digitalis as indicated. Pulmonary
rehabilitation.

Bronchogenic Carcinoma / Lung Cancer
Lung Cancer or Bronchogenic Carcinoma is malignant tumor(s)
resulting from damage to the genetic DNA of cells & mutations in
stem cell lines of the lungs. A tumor, neoplasm, or adenoma can be
malignant (cancerous) or non-malignant. Carcinoma is cancer derived
from lining cells of organs. Most common form of bronchogenic
carcinoma is adenocarcinoma, representing 30 to 35% of all lung
cancers. Other types are small cell, large cell, & squamous cell
carcinoma. The etiology can be idiopathic, but often cancers are
linked to carcinogens. Approximately 85% of bronchogenic carcinoma
cases are caused by smoking. Other causative substances include:
arsenic, asbestos, benzo (a) pyrene, certain man-made mineral
fibers, ionizing radiation, mustard gas, radon, & vinyl chloride.
S/S: Depends on type of cancer & degree of obstruction. Can be
asymptomatic, esp. in early stages. Obstructive tumors can cause
dyspnea, cough, hemoptysis, chest pain, & frequent pulmonary
infections. Percussion: dullness over affected area. Breath sounds:
diminished &/or wheezes, rhonchi, rales.
Tests: PFT: varies by type & severity. X-ray: possible nodule(s),
infiltrates, opaque mass, atelectasis, & mediastinal shift.
Tx: Depends on type of cancer. Prevention, avoid smoking, avoid
exposure to carcinogens. Treat symptoms & complications,
supportive care. Radiation therapy, chemotherapy, surgical resection,
lobectomy, or pneumonectomy as indicated.

Cancer / Carcinoma – See Bronchogenic Carcinoma

Clubbing, Digital Clubbing or
Hypertropic Pulmonary Osteoroarthropathy (HPO)
Bulbous enlargement & chronic inflammation of the fingers & toes.
Most commonly associated w/ chronic hypoxemia, chronic bronchitis,
emphysema, recurrent pulmonary infections, & congenital heart
defects. The angle of the skin-nail is greater than 160 degrees.

Coccidioidosis (San Joaquin Fever, Valley Fever)
A fungal infection that can cause respiratory symptoms. Coccidioides
immitis fungal yeast is found in soil of SW region of North America.
Acquired by inhalation, usually during a dust storm.
S/S: Often asymptomatic, but can lead to serious pneumonia-type
infection, or meningitis. Chest pain, cough, hemoptysis, dyspnea.
Tests: X-ray: possible cavitation, infiltrates, bronchopneumonia.

Tx: Pulmonary hygiene. Antifungal medication. Treat complications.

COPD (Chronic Obstructive Pulmonary Disease)
In the past, "COPD" was used to encompass most obstructive lung disorders. Currently, the American Thoracic Society classifies COPD as: Emphysema, chronic bronchitis, chronic asthmatic bronchitis. "Obstructive Pulmonary Disease" is the broader term & includes: Emphysema, Bronchitis, Asthma, Bronchiectasis, Cystic Fibrosis, & Bronchopulmonary Dysplasia (BPD). Information is listed individually about these obstructive diseases. (Most other pulmonary diseases are classified as restrictive.)

Cor Pulmonale (Right Heart Failure)
Hypertrophy & thickening of the right heart ventricle. Caused by pulmonary hypertension, & associated w/ lung disease.
S/S/Tests: Usually asymptomatic until cor pulmonale is advanced. Chest pain, dyspnea, lightheadedness, fatigue, venous distention, peripheral edema, hypoxemia, cyanosis. Characteristic heart sounds. EKG changes specific to right heart. Heart function can be evaluated w/ echocardiography, radionuclide studies, & cardiac catheterization (increased PAP & PVR). X-ray: enlarged right ventricle. Risk of pulmonary embolism is increased.
Tx: Treat underlying lung disease, resolve pulmonary hypertension, & relieve right-sided heart failure. Oxygen. Diuretics. Anticoagulant to prevent pulmonary embolism.

Cryptococcosis
Fungal systemic & respiratory tract infection. Occurs mainly in meninges (brain & spinal cord), lungs & skin. Cryptococcus neoformans is a fungal yeast that grows in pigeon droppings, soil, fruits & vegetables. Found throughout the world. Infection acquired by inhalation. ↑ risk w/ immunosuppression, AIDS, sarcoidosis, & Hodgkin's disease.
S/S/Tests: Usually asymptomatic to mild illness, but meningitis can develop. Fever, cough, hemoptysis, dyspnea, & headache. Blood & spinal fluid are tested for antibodies to Cryptococcus. X-ray: possible lesion, cavitation & infiltrates.
Tx: Varies by severity of illness. Antifungal medication. Treat symptoms & complications.

Cystic Fibrosis (CF, Mucoviscidosis, Pancreatic Enzyme Deficiency)
CF is a hereditary disease of the autonomic nervous system caused by inheriting two defective mutations of the CF gene. CF causes the exocrine glands to produce excess & abnormal secretions, resulting in tissue & organ damage, especially in the lungs, digestive tract, pancreas, liver, & sweat glands. The dehydration & viscous secretions occur because of the disruption of control of production of

a protein that regulates the transport of sodium & chloride across cell membranes.

S/S/Tests: Cough w/ copious amounts of viscous mucopurulent mucus, chest congestion, digital clubbing, dyspnea, hypoxemia, cyanosis. ↑ sweat chloride level. Enlarged lymph nodes, blocked bile ducts in the liver, gallbladder obstruction, pancreatic obstruction, intestinal obstruction (meconium ileus), digestive problems, nutritional deficiencies, infertility. Delayed growth in adolescents. Frequent pulmonary infections, w/ ↑ incidence of bronchitis, pneumonia, bronchiectasis, atelectasis. Breath sounds: rhonchi, wheezes. PFT: obstructive disease pattern.

Tx: Aggressive bronchial hygiene, secretion mobilization & clearance. CPT, vibratory PEP, or vibratory vest. Bronchodilators, dornase alfa, acetylcysteine, corticosteroids, oxygen. Infection control, antibiotics. Identify & treat nutritional deficiencies. Pancreatic enzyme replacements. Gene therapy is experimental. Liver &/or lung transplant when other therapies fail. Life expectancy has improved over the decades; half of CF patients now live 33 years or longer. Most CF patients eventually die of respiratory failure & a smaller number die of heart failure or liver disease.

Drowning, Near-drowning

Submersion in liquid w/ suffocation due to aspiration of liquid or laryngospasm. Rapid absorption of water into alveoli & bloodstream. Can be freshwater (hypotonic), or salt water (hypertonic). Dry-drowning is asphyxia due to laryngospasm w/ complete airway obstruction.

S/S: Cough expelling mucoid water, unconsciousness, convulsions, pulmonary edema, hypervolemia, hypoxemia, acidosis, atelectasis, cardiac arrhythmias, cardiopulmonary arrest.

Tx: Secure patent airway. Prevent aspiration. Correct acidosis, administer NaHCO3, & steroids. Restore fluid & electrolyte balance. High-concentration oxygen. Cardiopulmonary resuscitation & mechanical ventilation as indicated.

Emphysema

Emphysema (pink puffer) is classified as COPD. Dilation, enlargement & destruction of the air spaces beyond terminal bronchioles. Three pathological types include: centrilobar (centriacinar), panlobular (panacinar) & bullous (subcutaneous). Panlobular emphysema is associated w/ heredity factors & alpha$_1$ antitrypsin deficiency (A$_1$AD), also called A$_1$AT deficiency. Subcutaneous emphysema is differentiated from other types by the presence of large air-filled spaces in the soft tissues, esp. in apical & anterior subpleural areas; crepitus can be felt & air bubbles are seen.

Emphysema etiology includes: Smoking, air pollutants, occupational exposure to inhaled irritants, aging, chronic bronchitis, & infection. ↑ airway resistance, w/ air-trapping.

S/S: Cough, mucus production, dyspnea, cyanosis, ↑ use of accessory muscles, pursed-lip breathing, labored breathing, orthopnea, barrel chest. Possible digital clubbing, weight loss, & engorged neck veins. Cor pulmonale can develop.

Tests: ABG: hypoxemia, hypercapnia, compensated respiratory acidosis & chronic respiratory failure. Breath sounds: diminished, rhonchi, wheezes. PFT: obstructive disease & decreased DLCO. X-ray: hyperinflation, ↑ A-P diameter, depressed diaphragm, bullae, blebs.

Tx: Stop smoking. Avoid pollutants. Low-concentration oxygen (with consideration for hypoxic drive breathing), bronchial hygiene, expectorants, bronchodilators, & steroids. Pulmonary rehabilitation. Antibiotics for infection. Diuretics & digitalis as indicated. Maintain good nutrition. When other treatments fail, surgical resection, lung volume reduction, or lung transplantation is considered. For A_1AD, treatment involves alpha$_1$ proteinase inhibitor augmentation.

Empyema

Presence of purulent fluid in the pleural space. Caused by infection & complications of other diseases including: pneumonia, pulmonary abscess, tuberculosis, or bronchiectasis. (Also see pleural effusion.)

S/S: Varies by severity of effusion. Dyspnea, tachypnea, chest pain, cough, hemoptysis, hypoxemia. Breath sounds: diminished, or crackles over affected area.

Tests: X-ray: radiopacity shows fluid accumulation, pleural thickening.

Tx: Chest tube drainage. Oxygen. Treat symptoms & complications.

Fibrosis, Pulmonary

Fibrosis is the replacement of the normal components of a structure w/ fiber-containing tissue. This process can occur in lungs, organs & other structures Pulmonary fibrosis can result due to age, disease, or complications of other medical conditions. Any dysfunction that promotes prolonged or repetitious lung injury can begin an inflammatory process & release of mediators, followed by fibrosis. Some causes include: ARDS, aspiration, infection, bronchiectasis, pneumonitis, sarcoidosis, oxygen toxicity, necrosis, & inhalation of noxious gases.

S/S: Dyspnea, hypoxemia, ↑ work of breathing, & decreased compliance. Breath sounds: rales over affected area.

Tests: X-ray: fibrotic changes, atelectasis, & mediastinal shift toward fibrotic area.

Tx: Supportive treatment for symptoms & complications. Oxygen. Avoid exposure to pulmonary irritants.

Flail Chest

Defined as two or more ribs fractured in more than one location, resulting in unstable area of the chest wall. Causing the unstable chest area to move in the opposite direction of the rest of thorax. Can be caused by chest trauma or impact injuries.

S/S/Tests: Paradoxical respirations w/ chest wall moving in on inspiration & out on expiration. Dyspnea, ↑ work of breathing, pain in affected area, hypoxemia. X-ray: rib fractures, atelectasis.

Tx: Stabilize chest wall. Oxygen for hypoxemia. Mechanical ventilation if indicated. Pain management.

Guillian-Barré Syndrome

Also known as Inflammatory Demyelinating Polyneuropathy. There is ascending (from feet upward) muscle paralysis due to simultaneous malfunction of many nerves throughout the body, including nerves of the autonomic nervous system. Can be acute or chronic, short-term illness or permanent. (There exists a rare variant form is a descending paralysis.) Polyneuropathy is idiopathic. Presumed cause is an autoimmune reaction. The body's immune system attacks the myelin sheath that surrounds the axon of nerves.

Occurs frequently in diabetics. Other possible causes include: heredity, hypothyroidism, liver failure, kidney failure, alcoholism, anemia, malnutrition, excess amounts of vitamin B6, & certain cancers. Also, causes of acute polyneuropathy include: infection, toxic substances, & certain drugs like anticonvulsants, antibiotics, sedatives, sulfonamides, & chemotherapy drugs.

S/S: Ascending muscle weakness beginning at the feet/legs, & progressively upward. Can rapidly progress to become medical emergency, although some mild cases resolve spontaneously over several months. Most often occurs after an infection, a flu-like illness, surgery, or an immunization. (Symptoms usually begin within 1 to 8 weeks). Pins-and-needles sensation, numbness, burning pain, & loss of sensation. Possible dysphagia & aspiration. Eventually can lead to muscle paralysis; including paralysis of the diaphragm & muscles of ventilation, causing respiratory failure.

Tests: Diagnostic procedures: electromyography, nerve conduction studies, & analysis of cerebrospinal fluid via lumbar puncture (↑ protein in CSF indicative). Also certain blood & urine lab tests (to rule out or detect diabetes, kidney failure, or thyroid disorder).

Tx: Monitor pulmonary mechanics, esp. VC & NIF. Treat symptoms & complications. Ventilatory support as indicated. Nutritional support. Physical therapy. Plasmapheresis, infusion of immune globulin, corticosteroids & immunosuppressants as indicated. Early treatment promotes faster improvement, although residual weakness can last 3 years.

Hemothorax

Presence of blood in the intrapleural space. Can occur from chest trauma, injury, puncture, or surgery.
S/S/Tests: Varies from asymptomatic to severe respiratory distress. Dyspnea, tachycardia, tachypnea. Percussion is dull over affected area. Breath sounds decreased or absent over affected area. Tracheal shift away from affected side.
Tx: Chest tube drainage. Oxygen & assisted ventilation as indicated. Treat cause of bleeding. Treat complications.

Histoplasmosis

Systemic fungal infection & respiratory tract infection. Histoplasma capsulatum fungal yeast is found in soil & bird excreta, mainly in Ohio & Mississippi valleys. Acquired by inhalation.
S/S/Tests: Varies from asymptomatic to fatal. Flu symptoms, cough, dyspnea, chest pain. Pneumonia possible. X-ray: miliary calcification.
Tx: Antifungal medication. Treat symptoms & complications.

Influenza

Influenza is commonly known as the flu, a contagious viral respiratory tract infection. Types of influenza virus are A, B, & C. Most often acquired by inhalation of airborne droplets; but also acquired by touching a contaminated surface, then touching the mouth or nose.
S/S: Flu symptoms, cough, chest congestion, headache, weakness. Can also lead to pneumonia, bronchitis, sinusitis, or laryngitis. Complications can be fatal.
Tx: Prophylactic w/ flu vaccine. Treat symptoms & complications. Antibiotics if secondary bacterial infection develops.

Interstitial Lung Disease

A disease in the interspace area of lung tissue. Multiple causative factors including complications from other diseases. Also see: ARDS, bronchiolitis, fibrosis, pneumonia, & pulmonary edema.

Ketoacidosis (Diabetic Ketoacidosis)

Metabolic acidosis associated w/ diabetes mellitus. Etiology includes: Insufficient insulin, production of keto acids & acetone that ↓ pH, high protein diet, or alcohol withdrawal (alcoholic ketoacidosis).
S/S/Tests: Severe hyperventilation (Kussmaul breathing) w/ acetone odor. Vomiting, dehydration, headache. ABG: metabolic acidosis.
Tx: Correct metabolic acidosis, administer bicarb. Regulate glucose & insulin level.

Kyphoscoliosis (Including Kyphosis & Scoliosis)

Kyphoscoliosis is a combination of kyphosis and scoliosis, & is the most severe of the 3. Kyphosis is abnormal anterior-posterior curvature of the spine. Scoliosis is abnormal lateral curvature of the spine. These spinal/thoracic deformities can restrict ventilation & ↑ WOB. Etiology is: idiopathic, congenital defect, neuromuscular disorder, or TB of the bone.

S/S/Tests: Visible spinal/thoracic deformity, dyspnea on exertion, visibly ↑ WOB, hypercarbia, hypoxemia, atelectasis.

Tx: Assure adequate ventilation, assisted w/ BIPAP, or full ventilatory support if indicated. Oxygen for hypoxemia. Pulmonary hygiene. Prevent pulmonary infections & atelectasis.

Legionnaire's Disease

Diffuse pneumonia caused by ingestion of the bacterial organism Legionella pneumophilia, found in contaminated water.

S/S/Tests: Fever, chills, cough, malaise, muscle aches, headache. Breath sounds: rhonchi, wheezes, rales. Serological testing can confirm presence of organism. X-ray: bilateral patchy infiltrates w/ possible consolidation.

Tx: Treat symptoms & complications. Antibiotics, pulmonary hygiene, oxygen. Disease can resolve spontaneously or cause respiratory failure.

Lou Gehrig's Disease – See Amyotrophic Lateral Sclerosis (ALS)

Lung Cancer/Adenoma/Neoplasm – See Bronchogenic Carcinoma

Myasthenia Gravis

Neuromuscular disorder causing weakness & quick fatigue of the muscles, characterized by the descending manner. Most frequently involves the face, throat, & muscles of ventilation. The cause is idiopathic, but autoimmune disorder suspected. Reduced transmission of nerve impulses at the neuromuscular junction. Possible link to excess acetylcholinesterase, or lack of acetylcholine.

S/S/Tests: Weakness & fatigue w/ exertion. Early signs include facial muscle weakness, & ocular muscle weakness, droopy eyelids & double-vision, dysphagia, & dysphasia. Can lead to respiratory failure. Diagnostic: Tensilon test; tensilon is a short-acting anticholinesterase drug. If tensilon worsens the patient's condition, it is a cholinergic crisis. If the patient's condition improves temporarily w/tensilon, it is a myasthenic crisis.

Tx: Monitor respiratory muscle strength with VC & NIF, & treat as necessary. Ventilatory support if indicated. Rest, plasmapheresis, immunosuppressants, corticosteroids, anticholinesterase therapy (pyridostigmine or neostigmine). Atropine for cholinergic crisis (due to excessive anticholinesterase therapy) Removal of thymus gland (thymectomy) if a thymoma is present.

Neuromuscular Disease – See Guillian-Barré Syndrome, Myasthenia Gravis, or Amyotrophic Lateral Sclerosis (ALS)

Obstructive Pulmonary Disease – See COPD

Oxygen Toxicity
Medical complications due to exposure to high-concentration oxygen over an extended period of time. Toxicity is based on reactivity of free radicals. Normally associated w/ 40% FiO_2 or higher. (In premature neonates, RLF is associated w/ PaO_2 over 100 mmHg.)
S/S: Tachypnea, substernal pain, cough, nausea/vomiting. Can cause atelectasis, ↓ lung compliance, pulmonary edema & hemorrhage, fibrosis, refractory hypoxemia, hyaline membrane disease (HMD) & respiratory distress syndrome. In premature neonates, it can cause retrolental fibroplasia (RLF) or retinopathy of prematurity (ROP).
Tx: Avoid use of excessive oxygen. Use CPAP/PEEP therapy to reduce oxygen while maintaining adequate PaO_2. Also see ARDS symptoms & treatment.

Pectus Carinatum (Pigeon Breast)
Chest abnormality w/ excessive protrusion of the sternum & ↑ A-P diameter of the chest. Can be caused by other illnesses. Usually asymptomatic.

Pectus Excavatum (Funnel Chest)
Chest abnormality w/ excessive concavity of the sternum. Congenital defect. Usually asymptomatic.

Pickwickian Syndrome (Obesity Hypoventilation)
Hypoventilation associated w/ obesity. Upper airway obstruction during sleep, CNS abnormality & obesity can all be causative factors.
S/S: Hypersomnolence, hypoventilation, hypoxemia, cyanosis, hypercapnia. Can lead to cor pulmonale.
Tx: Weight reduction. CPAP or BiPAP. Oxygen. Treat complications & improve ventilation. (Also see Sleep Apnea.)

Pleural Effusion
Presence of fluid in the pleural space due to multiple causes. Exudates are an inflammatory effusion due to pneumonia, pulmonary emboli, empyema, or other pulmonary infection. Transudates from plasma filtering from blood vessels as in heart failure or hypoproteinemia.
S/S/Tests: Varies by severity of effusion. Dyspnea, tachypnea, chest pain, cough, hemoptysis, hypoxemia. Breath sounds: decreased, or crackles over affected area. X-ray: reveals location of effusion w/ obliteration of costophrenic angle, radiopaque homogeneous mass, mediastinal shift away from affected area, & possible atelectasis.
Tx: Thoracentesis, chest tube drainage. Oxygen. Treat symptoms & complications.

Pleuritis (Pleurisy)

A disease causing inflammation of the pleura. The etiology includes: infection, irritation, & inflammation of the pleura caused by bacteria, virus, neoplasm, or autoimmune diseases like lupus.

S/S/Tests: Dyspnea & localized pleuritic chest pain that can begin suddenly. Pain worse on inspiration & w/ cough. Auscultation: pleural friction rub. X-ray: pleural thickening, pleural effusion, possible atelectasis. Can lead to pneumonia.

Tx: Treat underlying cause. Pulmonary hygiene, encourage deep breathing & cough while splinting. Nonsteroidal anti-inflammatory drugs. Antibiotics for bacterial infection.

Pneumoconiosis

Group of occupational diseases caused by prolonged exposure & inhalation of dust or chemical fumes. Leads to chronic inflammation of the lungs & interstitial fibrosis. Occupational diseases in this group include: asbestosis, beryliosis, coal miner's disease, & silicosis.

S/S: Varies by concentration of dust/fumes inhaled & exposure time. Ranges from asymptomatic to: cough, hemoptysis, dyspnea, hypoxemia, chest pain, pneumonia.

Tests: X-ray: infiltrates, haziness, possible nodules.

Tx: Prevention as prescribed by OSHA: wear dust mask, limit exposure to causative agent. Treat symptoms & complications. Pulmonary hygiene. Bronchodilators. Oxygen. No known cure.

Pneumonia

Pulmonary inflammation & infection. Causative factors include many types of microscopic organisms, bacteria, virus, or fungus. Infection occurs by inhalation or carried to lungs via bloodstream. Some of the many types of pneumonia are bacterial, viral, aspiration, interstitial, bronchopneumonia, pneumocystis carinii (PCP), TB, dust/chemical fume pneumoconiosis, & hypersensitivity allergic pneumonitis. One idiopathic pneumonia is cryptogenic organizing pneumonitis (also called idiopathic bronchiolitis obliterans w/ organizing pneumonia or BOOP). High risk people include: immunocompromised, very young, very old, post-operative, debilitated, bedridden, paralyzed, those on mechanical ventilation, unconscious, or malnourished.

S/S: Variable by type & severity of pneumonia. Chills, fever, cough w/ sputum production, dyspnea, chest pain, hypoxemia. Breath sounds: rhonchi, wheezes, crackles.

Tests: Sputum culture to identify type of pulmonary infection. X-ray: to identify location & extent of infection, radiopacity, infiltrates, cavitation, consolidation.

Tx: Prevention via vaccine. For pneumoconiosis & allergic pneumonitis, avoid causative agents. Tx varies by type & severity of pneumonia. Pulmonary hygiene, deep-breathing exercises, secretion clearance. Oxygen. Bronchodilators. Antipyretic. Anti-inflammatory agents, corticosteroid, steroid. Antibiotics for bacterial infection.

Pneumonitis
Inflammation of the lungs. See pneumonia.

Pneumothorax
Presence of air in the intrapleural space, or within other areas of the thorax. Can be spontaneous pneumothorax (idiopathic & not resulting from injury). Can be traumatic, resulting from: trauma, penetrating injury, lung puncture, rib fracture, rupture of a lung bleb, rupture of the chest wall. Tension pneumothorax is an acute medical emergency where air leaks into pleural space but cannot escape.
S/S: Immediate onset of pain, dyspnea, tachypnea, & tachycardia. Hypoxemia. ↑ WOB can result in respiratory failure. Breath sounds: very diminished or absent over affected area.
Tests: X-ray: hyperlucency & absent vascular markings at location of pneumothorax, atelectasis, & trachea deviated away from affected side. Small spontaneous pneumothorax can be asymptomatic.
Tx: Small spontaneous pneumothorax may resolve without treatment. Otherwise: thoracentesis, chest tube, oxygen. Mechanical ventilation for respiratory failure.

Psittacosis (Chlamydia Psittaci Infection, Parrot Fever, Ornithosis)
Intracellular parasite infection, resulting from inhalation of Chlamydia psittaci, a gram negative bacteria found in bird droppings.
S/S/Tests: Ranges from asymptomatic to systemic infection, severe pneumonia, & respiratory failure. Cough, hemoptysis, fever, chills, dyspnea, hypoxemia, tachypnea. Blood culture tests positive for chlamydia.
Tx: Avoid exposure. Treat symptoms & complications. Anti-infectives.

Pulmonary Abscess
Localized collection of purulent material in the lung w/subsequent inflammation & necrosis. Typically caused by aspiration of infectious organism. Treated w/ pulmonary hygiene & antibiotics.

Pulmonary Edema
Accumulation of excessive fluid in the alveoli & interstitium. Any abnormality that ↑ capillary permeability or decreases colloidal osmotic pressure can cause fluid accumulation in the pulmonary system. Can be cardiogenic or noncardiogenic. I.E.: hypervolemia, congestive heart failure, heart disease, MI, sepsis, ARDS, near-drowning, pneumonia, embolism, oxygen toxicity, smoke inhalation, renal failure, hypertension, & severe hypoxemia.
S/S: Dyspnea, hypoxemia, cyanosis, cough w/ pink frothy secretions, hyperventilation, tachycardia, orthopnea, & anxiety. BBS: rales.

Tests: X-ray: prominent vascular markings, diffuse fluffy infiltrates w/ butterfly pattern, possible cardiomegaly.
Tx: Diuretics, digitalis, morphine, high-concentration oxygen. Mechanical ventilation w/ PEEP therapy as indicated.

Pulmonary Embolism (PE), Thromboembolism

Sudden partial or complete blockage of pulmonary artery blood flow. Usually the blockage is a blood clot (thrombus). Other types of emboli are fat, amniotic fluid, cancerous tumor fragments, foreign materials, & air emboli from intravenous infusion. Often, PE forms in leg or pelvic vein (deep vein thrombosis, DVT), but can form in an arm vein or in right side of heart. Predisposing factors are: blood stasis, blood clotting disorder, prolonged bed rest, surgery, obesity, paralysis, stroke, myocardial infarction, fracture in leg, hip or pelvis, & cancer.
S/S: Depends on severity of obstruction. Dyspnea, hyperventilation, tachycardia, arrhythmias, chest pain, anxiety, hypoxemia, cyanosis, cough, hemoptysis, atelectasis. Breath sounds: rales, rhonchi, wheezes. Can cause pulmonary necrosis, cor pulmonale, heart failure, or respiratory failure.
Tests: Ventilation/perfusion scan (V/Q scan), pulmonary angiography, CT scan, ultrasound, & D-dimer blood test.
Tx: Prevention of blood stasis. Compression elastic stockings, leg exercises, ↑ activity, anticoagulants, thrombolytic therapy, & clot filter placed in vein. Oxygen. Mechanical ventilation if required.

Pulmonary Hypertension

A condition where blood pressure in the pulmonary arteries is abnormally high. Normal PAP systolic pressure is 15-25 mmHg. Normal PAP diastolic pressure is 5-15 mmHg. Two types include primary & secondary. Primary hypertension is idiopathic, but suspected due to spasms of muscle layer in pulmonary arteries. Secondary hypertension (much more common) can occur as a result of any disease or disorder that affects the lungs, heart, or pulmonary blood flow. (Pulmonary emboli, cardiogenic shock, heart failure, hypervolemia, COPD, ARDS, or pulmonary fibrosis.)
S/S: Dyspnea on exertion, light-headedness, chest pain, weakness, & peripheral edema. Can lead to PE or cor pulmonale.
Tests: Definitive diagnosis via pulmonary artery catheter.
Tx: Treat underlying disease. Treat complications. Oxygen. Vasodilators, nitric oxide, diuretics. Anticoagulant to prevent pulmonary embolism.

Respiratory Failure

Failure of the pulmonary system failure that falls into two classifications: hypoxemic respiratory failure (inability to maintain adequate oxygen delivery to tissues), or hypercapnic respiratory failure (inability to maintain normalized removal of carbon dioxide from tissues). This can be acute medical emergency or chronic condition. It's caused by medical complications & any disorder that affects the lungs.

S/S: Depends on degree of hypoxemia & degree of hypercapnia. Symptoms range from dyspnea, anxiety, & respiratory distress to unconsciousness.

Tests: Clinical analysis & ABG's. Lung function tests like NIF & VC to assess respiratory muscle strength.

Tx: Oxygen for hypoxemia. BiPAP or full ventilatory support for hypercapnia.

Restrictive Disease – See specific listings

Sarcoidosis
Disease characterized by granulomas, & inflammation in the lymph nodes, lungs, liver, eyes, skin, &/or other tissues. Usually occurs in ages 30 to 50 years old, & is idiopathic. Possible causes include a genetic predisposition, a hypersensitive response to environmental factors, or extreme immune response to infection.

S/S/Tests: Can be asymptomatic. Can cause: fever, cough, malaise, dyspnea, hypoxemia, skin rash, skin lesions, enlarged liver, spleen, or lymph glands. Spontaneous pneumothorax, pulmonary fibrosis, pulmonary hypertension, cor pulmonale, or organ failure can develop.

Tx: Treat symptoms & complications. Pulmonary hygiene. Steroids helpful for some. Transplant if organ failure develops.

Scoliosis - see Kyphoscoliosis

Silicosis
Respiratory disease caused by inhalation of silica dust , which leads to inflammation, interstitial fibrosis, & scarring of lung tissue. Silica is a naturally occurring crystal found in rock beds & sand. Pulmonary symptoms can be caused by acute exposure to large amount of silica, or long-term exposure to small amounts. Occupational workers at risk are: road & building construction workers, miners, metal workers, stone cutters, sand-blasters, glass workers.

S/S: Variable, depending on exposure time & amount. Dyspnea, chronic cough, pulmonary edema, progressive massive pulmonary fibrosis, & obliteration of normal lung structures.

Tx: No known cure. Prevention as prescribed by OSHA: wear dust mask, limit exposure to silica. Treat symptoms & complications.

Sleep Apnea, OSA (Obstructive Sleep Apnea) & Central Sleep Apnea
Sleep apnea syndromes are a group of sleep disorders including obstructive sleep apnea (OSA), central sleep apnea, & mixed sleep apnea. These are characterized by repeat episodes of breathing cessation of breathing. Apneic periods cause hypersomnia during the

day & other symptoms as listed below. Apneic periods range from a 3 seconds to 2 or more minutes. Apnea is considered significant if there are at least five episodes per hour, with apnea lasting longer than 10 seconds. The condition is severe if more than 15 apneic episodes occur per hour. Note: There are many other sleep disorders including: insomnia, narcolepsy (isolated sleep paralysis), nocturnal myoclonus (restless leg syndrome or RLS), night terrors, sleepwalking, nocturnal asthma, nocturnal COPD disturbances, neuromuscular related sleep disorders, & Pickwickian syndrome. The etiologies follow:

OSA (most common type) is caused by obstruction of the upper airway during sleep. Causes of obstruction include obesity, anatomical anomaly resulting in narrowed airway, enlarged tonsils or enlarged adenoids.

Central sleep apnea (less common) is a neurological disorder caused by dysfunction in the brain stem, brain tumor, Ondine's curse, or idiopathic central hypoventilation. The brain fails to signal the body to breathe.

Mixed sleep apnea (rare) is a combination of obstructive & central sleep disorders.

S/S: Symptoms that can occur in all three sleep disorders are: daytime hypersomnolence, fatigue, irritability, difficulty concentrating, slowed thought processes, confusion, headaches, & hallucinations. Also, arrhythmias & hypertension can occur w/ moderate to severe hypoxemia. Prolonged severe sleep apnea can result in MI, heart failure, & respiratory failure.

Tests: The diagnosis & type of sleep disorder is best confirmed w/ polysomnography in a sleep laboratory. Tests include: EEG, EKG, electromyogram (EMG), electrooculogram (EOG), pulse oximetry, chest & abdomen motion detectors, & airflow measurement sensor. In OSA, a distinctive sign is snoring, which does not often occur in central sleep apnea. The sleep partner or parent can also give history of noticeable episodes of irregular breathing, apnea, & snoring.

Tx: OSA therapy includes: weight reduction, various devices to reduce snoring & splint the airway open, avoiding alcohol, cigarettes, & sedatives at bedtime, CPAP, & oxygen. Surgery to remove excess airway tissue, remove tonsils, or correct airway abnormalities.

Central sleep apnea therapy includes: antidepressant drugs & oxygen. CPAP is beneficial in some patients, but not in others.

Mixed sleep apnea includes a combination of therapies.

Smoke Inhalation

Inhalation of smoke into the lungs. Fire, smoke, or fumes can cause pulmonary damage.

S/S: Depends on amount of smoke/fumes inhaled. Burns over face &/or body, dyspnea, hypoxemia. Respiratory failure can develop.

Tests: Carboxyhemoglobin (HbCO) greater than 5% causes adverse effects.

33

Tx: Hyperbaric oxygen therapy (HBO) if available, or 100% oxygen. Mechanical ventilation for respiratory failure. Steroids.

Thrush
Fungal infection that can occur in the oral cavity. Caused by Candida albicans. Prevented by rinsing mouth after using inhaled steroids.

Tuberculosis (TB or MTB, Mycobacterium Tuberculosis)
Serious bacterial infection that usually affects the pulmonary system, but can also affect other parts of the body (extrapulmonary TB). Pulmonary tuberculosis results in inflammatory reaction, tubercles, & necrotizing lesion w/ caseating granuloma in the center. Fibrotic tissue eventually replaces much of the granulomatous lesion. Miliary TB is disseminated form of TB, where the infection spreads through the lymphatics & bloodstream to other organs. TB infection is caused by inhaling airborne droplet nuclei of Mycobacterium tuberculosis.
S/S: Can be asymptomatic indefinitely in the latent phase. Cough w/ mucopurulent sputum, hemoptysis, dull chest pain, dyspnea, tachycardia, fatigue, weight loss, night sweats, fever, & enlarged lymph nodes. High risk groups includes: immunocompromised, impoverished, debilitated, diabetic, & malnourished.
Tests: Positive PPD skin test. Sputum sample- positive acid-fast stain. X-ray: infiltrates, consolidation, & cavitation (usually in apical segment).
Tx: Treatment consists of long-term drug therapy w/ antibiotics/ antitubuculotics for 6-9 months. Drugs include isoniazid (INH), ethambutol, rifampin, & streptomycin. Treat complications & relieve symptoms. Pulmonary hygiene. If left untreated, active TB can seriously damage lungs or other organs & can be fatal. Assure respiratory isolation to protect others.

Chapter 3

MICROBES: Bacterial, Viral, Fungal & Bioterror Agents

ALSO SEE: Pulmonary Diseases – Chapter 2

Bacterial Organisms & Disease

NOTES:
G+ Gram positive
G– Gram negative
Some of the bacterial organisms most commonly seen in the clinical setting are listed first in the highlighted section of this table.

Bacterial Organisms (most commonly seen)	Diseases Caused by Bacterial Organism
Diplococcus pneumoniae	See Streptococcus.
Enterococcus faecium or VRE (vancomycin-resistant enterococcus)	VRE blood infection, pneumonia, very serious nosocomial infection.
Escherichia coli (G–)	Pneumonia, cystitis. Habitat- intestinal tract.
MRSA	See Staphylococcus.
Mycobacterium tuberculosis (G+)	Tuberculosis (TB or MTB), infection of lungs & other organs.
Mycoplasma pneumoniae (G–)	Atypical pneumonia. Habitat- sputum, blood, GI tract.
Pneumococcus pneumoniae	See Streptococcus.
Pseudomonas aeruginosa (G–)	Nosocomial resp infection, pneumonia. Habitat- colon, skin, water, soil, & respiratory equipment.
Staphylococcus aureus & MRSA (methicillin-resistant staph aureus) (G+)	Nosocomial infection, pneumonia, empyema, meningitis septicemia. Habitat- respiratory tract, skin, GI tract.
Streptococcus pneumoniae Pneumococcus pneumoniae Diplococcus pneumoniae (G+) (all 3 are the same bacteria)	Pneumonia, empyema, meningitis, sinusitis, otitis. Strep species is also associated w/ strep throat, endocarditis, skin lesions, & UTI.
VRE	See Enterococcus.

Bacterial Organisms (continued)	Diseases Caused by Bacterial Organism
Bacillus anthracis (G+)	Anthrax infection- 3 types are cutaneous, ingested, & inhaled.
Bordetella pertussis (G–)	Whooping cough, nasopharyngeal infection.
Chlamydia psittaci (G–)	Psittacosis, parrot fever, ornithosis, pneumonia.
Clostridium botulinum (G+)	Botulism- food borne, wound infection, or infantile. Causes paralysis. Habitat- soil & feces.
Clostridium perfringens (G+)	Gas gangrene. Habitat- soil & feces.
Clostridium tetani (G+)	Tetanus, lock jaw, rigid muscle paralysis, respiratory failure. Bloodstream infection via infected wound (rusty nail, splinter, animal or insect bite). Habitat- dirt, soil, & feces.
Corynebacterium diptheriae (G+)	Diphtheria, airway obstruction, pharynx & larynx mucosal necrosis, skin infections.
Coxiella burnetii (parasitic bacteria-like organism)	Q-fever, Rocky Mountain Fever, pneumonia, rickettsia disease. Habitat- ticks, mites & lice.
Friedlander's bacillus	See Klebsiella pneumoniae.
Haemophilis influenzae (G–) (different than viral flu influenza)	Epiglottitis, laryngitis, sinusitis, nasopharyngitis, meningitis, respiratory infection, pneumonia, bronchitis.
Klebsiella pneumoniae, (G–) or Friedlander's bacillus	Necrotizing pneumonia, lung abscess, respiratory infection, UTI, septicemia, endocarditis.
Legionella pneumophilia (G–)	Legionnaires disease, diffuse pneumonia, pneumonitis. Habitat- contaminated water.
Meningococcal meningitis or Neisseria mengitidis	Meningitis, serious infection & inflammation of the brain & spinal cord meninges. Spread by respiratory droplets. Meningitis is also caused by other bacteria, viruses, & diseases.
Proteus vulgaris (G–)	Pneumonia, bacteremia, UTI, gastroenteritis.

Salmonella enteritidis & Salmonella typhimurium	Salmonellosis- food poisoning due to ingestion of contaminated food or contact w/ infected animal feces or reptiles
Serratia marcescens (G–)	Serious pneumonia, empyema, septicemia, wound infections.

Viral Organisms & Disease

Viral Organism	Disease Caused by Viral Organism
Cytomegalovirus (CMV)	CMV infections, Giant Cell Inclusion Disease (CID)
Epstein-Barr virus	Mononucleosis
Hepatitis A,B,C virus	Hepatitis
Human immunodeficiency virus (HIV)	AIDS; profound immunosuppression & opportunistic infections like PCP, TB, Kaposi's sarcoma, lymphoma, & others
Influenza	Common Cold & Common Flu
Measles virus	Measles/ German Measles/Rubella
Mumps virus	Mumps
Respiratory syncytial virus (RSV)	Bronchiolitis
RSV, adenovirus, or parainfluenza virus	Croup /LTB
RNA virus	Poliomyelitis
Rabies virus, rhabdovirus Lyssavirus	Rabies, encephalitis, CNS disease
Human herpesvirus HHV-6 (& possibly HHV-7)	Roseola infantum, Sixth disease
Varicella virus	Chickenpox
Variola virus	Smallpox
Viruses of many different types can also cause:	Viral Encephalitis or Meningitis (Note: meningitis can be caused by virus, bacteria, or disease)

Fungal Organisms & Disease

Fungal Organism	Diseases Caused by Fungal Organism
Actinomyces fungus group	Actinomycosis
Aspergillus fumigatus	Aspergillosis
Blastomyces dermatitidis	Blastomycosis
Candida albicans	Thrush
Coccidioides immitis	Coccidioidosis, San Joaquin Fever, or Valley Fever
Cryptococcus neoformans	Cryptococcosis
Histoplasma capsulatum	Histoplasmosis
Pneumocystis carinii (classified as protozoan or fungus)	Pneumocystis carinii pneumonia (PCP)

Potential Bioterror Agents

The following is a list of potentially dangerous boiwarfare agents if utilized as weapons of terrorism. For the most current updates, and comprehensive information, please visit the U.S. government – CDC website at: **www.cdc.gov**

Bacillus anthracis (Anthrax)
Brucellosis (Mediterranean or Malta Fever)
Clostridium botulinum toxin (Botulism)
Francisella tularensis (Tularemia)
Staphylococcal Enterotoxin B (SEB)
Variola virus (Smallpox)
Venezuelan equine encephalitis virus (VEE)
Yersinia pestis (Plague)
Nerve Agents: cyanide, sarin, & skin-blistering compounds

Adverse effects depend on the type of microbe, the amount exposed to, & the route & duration of exposure. Symptoms can include: fever, headache, malaise, nausea, vomiting, diarrhea, cough, dyspnea, respiratory distress, & respiratory failure. Adverse effects range from minor skin irritations to complete incapacitation & death. Usually the percutaneous & oral routes are less severe in terms of casualties & fatalities than the respiratory route.

Respiratory route of exposure:
When used as biowarfare agents, many microbes & toxins can be dispersed by aerosol, and can remain suspended for hours. Aerosol with particles of 1 to 10 microns can penetrate the lungs, distal bronchioles & terminal alveoli. The aerosol is colorless, odorless, & invisible.

Other routes of exposure are:
percutaneous (by contact with exposed skin)
ingestion (by consumption of contaminated food or water)

Personal Protection:
It is advisable for all healthcare personnel who are treating suspected cases to utilize Personal Protective Equipment (PPE) including: gloves, isolation gowns, & high filtration masks. Thorough handwashing with proper technique is also important.

NOTES:

Chapter 4 RESPIRATORY MEDICATIONS

Definitions

Drug: any chemical that alters an organism's functions or processes.
Pharmacokinetics: the study of absorption, metabolism, & excretion of drugs in the body.
Metabolism: a chemical process carried on within the body to maintain life. Drugs are absorbed most rapidly from large surface areas such as the lungs, or by IV. The liver is the major site of metabolism. After being metabolized, the drug is eliminated from the body primarily by the kidneys.
Tachyphylaxis: rapid development of drug tolerance, requiring higher doses to achieve the same results.

Notes

Most current drug information: See the US Food & Drug Administration website at: (www.fda.gov). This is useful as an up-to-date reference & to get the facts about new drugs that continuously become available. Also note that indications, contraindications, & safe dosage amount can change over time.

A contraindication for any drug: is hypersensitivity or allergy to the drug. Children & elderly generally require reduced dosages

5 R's to Assure Correct Drug Administration

Right Patient. Right Medication. Right Dose.
Right Route. Right Time.

Administration Routes for Pulmonary Medications

Inhalation:
Advantages – direct delivery to lungs, rapid onset of action, & reduced systemic side effects. Methods – nebulizer, MDI, or DPI.

Oral:
Advantages – certain oral drugs can have longer duration of action than the inhaled drugs. Disadvantage – slower onset of action than the inhaled route. Methods – tablet, capsule, or pill.

Intravenous & Intratracheal:
Advantages – Very rapid onset of action, very useful in emergencies.

SQ & IM:
Subcutaneous & Intramuscular routes are useful for delivery of many drugs given by nurses. Not typical routes for pulmonary medications.

Calculating Drug Dosages

When calculating active ingredient of a drug from percent strength solutions, remember that % solution represents grams of drug per100 mL of solution.

Formula:
Multiply the % strength by 10. Then, multiply the result by the number of mL. This will give you the mg of active ingredient.

(% strength x 10) x number of mL = mg of active ingredient

1 mL of 1% = 10 mg   2 mL of 1% = 20 mg
1 mL of 5% = 50 mg   2 mL of 5% = 100 mg
1 mL of 10% = 100 mg   2 mL of 10% = 200 mg

Ratio Formulas:

1:100 = 1% solution   (1 / 100 = .01 = 1%)
1:50 = 2% solution   (1 / 50 = .02 = 2%
1:20 = 5% solution   (1 / 20 = .05 = 5%)

Remember: 1 mL = 1 cc or 20 drops

Calculating Pediatric Estimated Safe Dosages:

Refer to employers policy, manufacturer's literature, an approved drug reference book when giving pediatric dosages.

"Clark's Rule" is one method to calculate estimated safe dosage (ESD) for children. Clark's rule is based on reducing dosage by weight, and using 150 pounds as the average weight of an adult in the general population.
Clark's Rule for ESD:
 (childs weight in pounds x average adult dose) / 150
 Example: (60 pounds x 3 mg adult dose) / 150 = 1.2 mg ESD

BRONCHODILATORS

Bronchodilators affect the autonomic nervous system.
The autonomic nervous system is divided into 2 branches:

Sympathetic Branch & Parasympathetic Branch
(Table A & B) **(Table C)**

Sympathetic branch drugs
are also known as Sympathomimetic or Adrenergic.

See Tables A & B. Receptors in the sympathetic branch are known as Alpha, Beta 1 (B1), & Beta 2 (B2). Both B1 & B2 drugs are called beta agonists. The drugs work by stimulating one or more of these receptors. The sympathetic branch also contains the "fight or flight" response & directly innervates the smooth muscles.
  Alpha response results in increased blood pressure & vasoconstriction of arteriolar smooth muscles.
  Beta 1 (B1) response results in increased heart rate & increased strength of cardiac contractions.
  Beta 2 (B2) response results in direct bronchodilation of the lungs, stimulates mucociliary activity & has a weak vasodilating effect. The goal of a bronchodilator is to stimulate B2 without the adverse effect of increased heart rate. However, slight B1 stimulation can be a side effect of B2 drugs.

Parasympathetic branch drugs / Anticholinergics
are also known as Parasympatholytic or Antimuscarinic.

See Table C. Parasympathetic stimulation results in bronchoconstriction & decreased heart rate, which is the opposite effect of sympathetic stimulation. The goal of bronchodilation via this branch is to **block** parasympathetic stimulation, thereby blocking the mechanisms that cause bronchoconstriction, & indirectly resulting in bronchodilation. So these drugs are also called "back door bronchodilators".
The mechanism of action is to inhibit vagally-mediated reflexes (vagally-induced bronchospasm & bradycardia) by antagonizing acetylcholine at muscarinic receptors in the lungs. Anticholinergics have a slower onset of action than B2 bronchodilators, but a longer duration of action. These are not used in emergencies, they are used as maintenance therapy. Anticholinergics can be mixed with B2 bronchodilators to achieve bronchodilation by stimulating the sympathetic branch, while blocking the parasympathetic branch.

BRONCHODILATORS

Sympathomimetic, Adrenergic
Beta 2 Specific & Combination Drugs (shaded)

Generic Name Brand Names	Routes of Administration	Normal Adult Dosage	Duration of Action	Onset of Action
arformoterol Brovana*	Nebulizer	15 Mcg dose BID	12 Hr	7 min
albuterol Aerolin Proventil Respolin Salbutamol Ventolin Volmax	Nebulizer............ MDI (90mcg/puff) DPI (200mcg/cap) Tablets (2 or 4 mg) Syrup (2mg/5mL)	2.5 - 5 mg/NS Q 4-6 Hr... 1 - 3 puffs Q 4-6 Hr......... 200 mcg Q 4-6 Hr............ 2 - 4 mg Q 6-12 Hr.......... 2 - 4 mg Q 6 Hr..............	..4 - 6 Hr ..4 - 6 Hr ..4 - 6 Hr ..6 -12 Hr ....6 Hr...	...5 min.... ...5 min... ...5 min... ...30 min.. ...30 min..
Accuneb (pediatric dose albuterol)	*Accuneb Nebulizer 0.63 mg or 1.25 mg in NS- 3mL vial*	.... *Accuneb is Pediatric dose age 2-12:* *One vial TID or QID*	.... *4 - 6 Hr..*	.... *5 min*
albuterol & ipratropium bromide "albuterol & Atrovent" **Combivent** .... **Duoneb**	MDI 2 puffs Q6 Hr or QID. Max 12 puffs daily. 1 puff contains 90 mcg albuterol & 18 mcg ipratropium. .... Nebulizer 1 vial Q6 Hr or QID. 3 mL vial contains 3 mg albuterol & 0.5 mg Atrovent in NS.		albuterol 4 - 6 Hr Atrovent 5 - 8 Hr	albuterol 5 min Atrovent 20 min
levalbuterol Xopenex	Nebulizer................	0.63-1.25mg/NS Q 4-8 Hr	5 - 8 Hr...	5 -10 min
bitolterol Tornalate	Nebulizer................ MDI (0.37mg/puff)..	2.5 mg Q 6-8 Hr 2 puffs Q 6-8 Hr	6 - 8 Hr...	4 min
pirbuterol Maxair	MDI (0.2 mg/puff)...	2 puffs Q 4-6 Hr	5 - 6 Hr...	5 min
terbutaline Brethaire Brethine Bricanyl	MDI (0.2 mg/puff)... Tablets.................... Subcutaneous..........	2 puffs Q4-6 Hr................ 2.5 to 5 mg TID............... 0.25mg Q 4-8 Hr..............	4 - 8 Hr... 6 - 8 Hr... 4 - 8 Hr...	....5 min... ...30 min.. ...10 min..
salmeterol Serevent*	MDI (25 mcg/puff).. DPI (50 mcg/puff)...	2 puffs BID 1 puff BID	12 Hr....	10-20 min

43

salmeterol & fluticasone "Serevent & Flovent" **Advair***	DPI Diskus 1 puff BID. 1 puff contains 100, 250, or 500 mcg Flovent, plus 50 mcg Serevent. Labeled as 100/50, 250/50, or 500/50.	Serevent 12 Hr Flovent- varies	Serevent 10-20 min Flovent- varies

Table A Notes: Albuterol, levalbuterol, & salmeterol are the most common direct bronchodilators with B2 receptor specificity. These drugs in are differentiated by the time for onset of action, & the duration of action. Combination drugs containing B2 specific drugs are also included.
*Brovana, Serevent & Advair are not recommended for relief of acute episode of bronchospasms. Rinse mouth after use of Advair to avoid thrush.

Table B. OTHER BRONCHODILATORS

Sympathomimetic, Adrenergic
Non-specific for Beta 2

Generic Name Brand Names (& Receptors*)	Routes of Administration	Normal Adult Dosage	Duration of Action	Onset of Action
epinephrine# Adrenalin Bronkaid (A++,B1++, B2++)	I.V. or Intratracheal Nebulizer (1%)........	0.5 to 1 mg & repeat PRN 2.5 to 5 mg PRN	1-3 Hr	1 – 2 min
isoetherine Bronkosol Bronkometer (B1+, B2++)	Nebulizer................ MDI 340 mcg/puff..	2.5 to 5 mg Q 4-6 Hr 2 puffs QID	4 Hr	5 min
isoproterenol Isuprel (B1++, B2++)	Nebulizer (0.5%)	0.25 to 0.5 mL +NS QID Max 5 doses daily	1 - 2 Hr	1 - 5 min
metaproterenol Metaprel Alupent (B1+, B2++)	Nebulizer................ MDI (0.65mg/puff).. Tablets...................	15 mg Q4 Hr.................... 2-3 puffs Q4 Hr............... 10-20 mg Q6-8 Hr..........	3 - 4 Hr 3 - 4 Hr 6 - 8 Hr	5 min 5 min 30 min
racemic epinephrine## Vaponefrin MicoNephrin (A+,B1+, B2+)	Nebulizer (2.25%)	0.25 to 0.5 mL +NS PRN	1 - 4 Hr	10 min

Table B Notes: These drugs are less specific for B2, & more specific for Alpha or B1 receptors. These drugs are sometimes indicated primarily for purposes

44

other than bronchodilation, or in addition to bronchodilation (i.e. to increase heart rate or to increase blood pressure).

#Epinephrine is used primarily in emergency situations, such as status asthmaticus & cardiopulmonary arrest (to increase HR & BP, & bronchodilate).
##Racemic epinephrine is used primarily for the alpha effects of reducing edema in the upper airway, reducing stridor, (i.e. post-extubation)& treating croup.

*Receptors - Describes the specificity of the drug for certain receptors.
- A= Alpha (increases blood pressure)
- B1= Beta 1 (increases heart rate)
- B2= Beta 2 (direct bronchodilation)
- ++ Strong stimulation of receptor
- + Moderate stimulation of receptor.

- **Indications for Bronchodilators:** Prevention & treatment of bronchospasm or reversible airflow obstruction. To decrease dyspnea, decreased WOB, improve breath sounds, improve air movement, improve flow rates, & improve exercise tolerance.

- **Contraindications for Bronchodilators:** Tachycardia or arrhythmias. Use caution in hypertension, coronary artery disease, hyperthyroidism, & diabetes. Note: beta-blocker drugs including propranolol, can block bronchodilating action of adrenergic drugs.

- **Adverse Reactions to Bronchodilators:** The following adverse reactions are more common in drugs that are less specific for Beta 2 receptors. Side effects include: tachycardia, heart palpitations, nervousness, tremors, headache, vertigo, diaphoresis, nausea, & hypokalemia. Tachyphylaxis & paradoxical bronchospasms possible with excessive use.

Table C. ANTICHOLINERGIC BRONCHODILATORS

also known as: **Parasympatholytic or Antimuscarinic**

Generic Name Brand Names	Routes of Administration	Normal Adult Dosage	Duration of Action	Onset of Action
atropine sulfate (strong B1, used primarily to increase heart rate)	IV or Intratracheal	0.5-2 mg PRN	1 - 4 Hr	5 min
	Nebulizer	1 mg QID	1 - 4 Hr	5-20 min
ipratropium bromide Atrovent	Nebulizer(0.02% sol)	0.5 mg/NS Q6 Hr or QID	5 - 8 Hr	20 min
	MDI 18 mcg/puff	2 puffs Q6 Hr or QID		
tiotropium bromide Spiriva	DPI 18 mcg/capsule	1 capsule QDay	24 Hr	20 min

(See anticholinergic bronchodilator notes earlier in this chapter.)

- **Indications for Anticholinergics:** Maintenance treatment for bronchospasms associated with COPD & certain cases of asthma. Anticholinergics only effective if bronchoconstriction is due to cholinergic activity.

- **Contraindications for Anticholinergics:** Not indicated for acute bronchospasms when rapid response is required, use Beta 2 drug instead. .Hypersensitivity to atropine, soy products, & peanuts.

- **Adverse Reactions to Anticholinergics:** nausea, nervousness, cough, dry mouth, thick secretions, tachycardia, palpitations, vertigo, headache, blurred vision.

Table D.   **CROMOLYN & NEDOCROMIL**

Generic Name Brand Names	Routes of Administration	Normal Dose	Duration of Action	Onset of Action
cromolyn sodium Intal Crolom	Nebulizer 20 mg/2mL NS.. MDI 800 mcg/puff............. Spinhaler 20 mg/Capsule.....	(Age 4 & older) 20 mg QID 2 puffs QID 20 mg QID	2 - 6 Hr	20-30 min
nedocromil sodium Tilade	MDI 1.75 mg/puff	(Age 12 & older) 2 puffs QID	3.5 Hr	20-30 min

Table D Notes: Cromolyn sodium & nedocromil sodium are nonsteroidal anti-inflammatory drugs (NSAID). These drugs work by inhibiting mast cell degranulation during antigen-antibody reaction, thus preventing the release of chemical mediators that cause inflammation & bronchospasm including: histamine, serotonin, & slow-reacting substance of anaphylaxis (SRS-A). Cromolyn & nedocromil sodium have proven to be more beneficial in children with asthma than in adults.

- **Indications for cromolyn & nedocromil sodium:** Prophylaxis & maintenance treatment of mild to moderate bronchial asthma.

- **Contraindications for cromolyn & nedocromil sodium:** Not for acute asthmatic episodes. Not for acute bronchospasms.

- **Adverse Reactions to cromolyn & nedocromil sodium:** nausea, pharyngitis, cough, bronchospasm, dyspnea headache, dizziness, rash.

Table E.   LEUKOTRIENE INHIBITORS

Generic Name Brand Names	Route of Administration	Normal Adult Dosage	Duration of Action	Onset of Action
montelukast sodium Singulair	PO - Tablet	4 - 10 mg QHS	24 Hr	1 - 4 Hr
zafirlukast Accolate	PO - Tablet	20 mg BID	12 Hr	1 - 3 Hr
zileuton Zyflo	PO - Tablet	600 mg QID	6 - 8 Hr	1 - 2 Hr

Table E Notes: The mechanism of action in leukotriene inhibitors is to block the leukotrienes that cause inflammation & bronchoconstriction. This may enable patients to reduce dependence on steroidal medications.

- **Indications for Leukotriene Inhibitors:** Prophylaxis & long-term maintenance treatment of asthma.

- **Contraindications for Leukotriene Inhibitors:** Not for acute bronchospasms. Not for acute asthmatic episodes.

- **Adverse Reactions to Leukotriene Inhibitors:** nausea, vomiting, diarrhea, headache, muscle aches, weakness, liver inflammation, low white blood cell count, infection, fever.

Table F.   XANTHINE DRUGS

Generic Name Brand Names	Routes of Administration	Normal Adult Dosage	Duration of Action	Onset of Action
aminophylline Phyllocontin Truphylline	IV Emergency...... Maintenance....	4.7 to 7 mg/kg f/b maint 0.5 to 0.9 mg/kg/hr	Varies	5-15 min
theophylline Theo-Dur, Theolair, Theovent, SloBid, Aerolate, Bronkodyl, Elixophyllin, Uniphyl, Quibron-T	PO Tablet or Liquid	Max daily dose is lesser of 13 mg/kg or 900 mg. Avg 400 mg/day divided into even doses Q6-8 Hr. (Extended release Tabs Q8, 12, or 24 Hr)	Varies	15-60 min

Table F Notes: Xanthine drugs are also know as methylxanthine drugs & phosphodiesterase inhibitors. The mechanism of action is to inhibit phosphodiesterase, the enzyme that degrades cyclic AMP. This results in relaxation of the bronchial airways & pulmonary vasodilation.

- **Indications for Xanthine Drugs:** Symptomatic relief of bronchoconstriction.

- **Contraindications for Xanthine Drugs:** Active peptic ulcer, or seizure disorder (unless treated w/anticonvulsants). Hypersensitivity to xanthine compounds (including caffeine). Use caution if cardiac arrhythmias or tachycardia exists.

- **Adverse Reactions to Xanthine Drugs:** nausea, vomiting, diarrhea, diuresis, stomach pain. palpitations, tachycardia, arrhythmias, hypotension. CNS stimulation, dizziness, seizures, insomnia, nervousness, irritability. (To minimize side effects & maximize effectiveness, monitor & maintain therapeutic blood theophylline level at 10 to 20 mcg/mL. Over 20 mcg/mL is toxic.

Table G.   MUCOLYTICS & WETTING AGENTS

Generic Name Brand Names	Normal Adult Dosage	Notes
acetylcysteine Mucomyst	2 to 4 cc of 10% or 20% solution PRN TID or QID by Neb	Decreases viscosity & breaks disulfide bonds in mucus. Give w/ bronchodilator to prevent bronchospasm. After opening: store in refrigerator & use within 4 days or dispose. (acetylcysteine is also an antidote for acetaminophen OD)
dornase alfa PulmoZyme	2.5 cc of 1mg/ml solution QDay PRN Use approved Neb like Hudson Updraft or Marquest Acorn II	Hydrolyzes sputum DNA, & decreases mucus viscosity & elasticity. Used to treat CF & other infection/ disease w/ purulent mucus. Can cause pharyngitis & cough. Currently not compatible to mix w/other drugs in Neb.
saline 3% hypertonic NaCL	2-5 cc by Neb PRN	Lung irritant utilized to induce sputum. Can cause bronchospasms; may need bronchodilator. Use caution in those w/edema or salt restrictions.
saline 0.9% NS isotonic NaCL	2-5 cc by Neb or Instill Intratracheal PRN	Humidifies & thins mucus. Used as a drug dilutent for bronchodilators. Can cause bronchospasm if given alone.
saline 0.45% hypotonic NaCL	2-5 cc usually by ultrasonic Neb PRN	Humidifies & thins mucus. Deposits more distally, & less irritating to mucosa than 0.9% NS.
sodium bicarbonate NaHCO$_3$	2 - 5 cc of 2% to 7.5% solution by Neb, or Instill Intratracheal PRN	Breaks saccharide chains in mucus, decreases adhesiveness, & lowers surface tension. Give bronchodilator first to prevent bronchospasm. Caution in edema, salt restrictions, or metabolic alkalosis.

Table G Notes: Mucolytics are utilized to thin, liquefy, & reduce viscosity thick, tenacious secretions. Facilitates the production & removal of respiratory tract fluids by expectoration or suction. Many mucolytics can cause bronchospasms, which can be minimized or avoided by giving a bronchodilator, just prior to mucolytic, or with mucolytic (consider compatibility issues).

Table H. CORTICOSTEROID ANTI-INFLAMMATORY DRUGS

Inhaled Corticosteroids

Generic Name Brand Names	Routes of Administration	*Normal Adult Dosage	Duration of Action	**Onset of Action
beclomethasone Beclovent Rotacaps, Beclodisk, Vanceril .... QVAR HFA	MDI/DPI 42-50 mcg/puff .... DPI 40-80 mcg/puff	2 puffs TID or QID, or 2-4 puffs BID (max 20 puffs/day) .... 1-2 puffs BID	Varies	Varies
budesonide Pulmicort	DPI 200 mcg/puff... Nebulizer.................	1-3 puffs BID 0.5 mg/NS BID	Varies	Within 24 Hr
dexamethasone Decadron	MDI 84 mcg/puff	2-3 puffs BID or QID	2-3 days	1 hour
flunisolide Aerobid .... Aerospan HFA	MDI 250 mcg/puff .... MDI 80 mcg/puff	2 puffs BID (max 8 puffs/day) .... 2 puffs BID	Varies	Varies
fluticasone Flovent (also see Advair in Table A)	MDI 44,110, or 220 mcg/puff DPI 50, 100, or 250 mcg/puff	2 puffs BID (max 1000 mcg/day)	2-3 days	Within 24 Hr
triamcinolone Azmacort	MDI 100 mcg/puff	2 –4 puffs BID to QID (max 16 puffs/day)	Varies	Varies

Systemic Corticosteroids

hydrocortisone Solu-cortef	PO – Tablet............ IV..........................	5-30 mg BID-QID 4 mg/kg Q4-6 Hr	Varies	Varies
methylprednisolone Medrol Solu-medrol	PO – Tablet............ IV..........................	4-48 mg Q day.......... 1-2 mg/kg Q4-6 Hr.....	1-3 Hr.... Rapid......	1-2 days 7 days
prednisone Prednisone, Deltasone, Meticorten, Orasone, Sterapred	PO – Tablet or Liquid	5-60 mg Q day	Varies	Varies

Corticosteroid drugs pending FDA approval:
ciclesonide (Alvesco) & roflumilast (Daxas)

Table H Notes: Corticosteroids are also known as glucocorticoids. Natural corticosteroids are endogenous hormones produced in the adrenal cortex. Corticosteroid drugs are used to treat inflammatory pulmonary diseases including asthma & COPD. These drugs reduce inflammation by stabilizing leukocyte lysosomal membranes & they may increase the effectiveness of B2 bronchodilators. Corticosteroids also suppress immune responses, including response of pulmonary mucosa to allergens. Other routes of administration

include nasal & intramuscular. Instruct patients to rinse mouth after corticosteroid inhalers to prevent thrush.
* The dosage & medication may need to be adjusted for patients previously taking other corticosteroids. Consider drug interactions.
**Corticosteroids can take 1-4 weeks to reach full effectiveness.

- **Indications for Corticosteroids:** Prophylaxis & long-term maintenance therapy of inflammatory lung diseases. Intranasal steroids are used to control seasonal allergic & non-allergic rhinitis.
Inhaled corticosteroids have lower incidence & lower severity of adverse effects than systemic corticosteroids. The inhaled route delivers medication directly to the lungs using a smaller dose, & only a small portion of the drug is absorbed into the bloodstream.
Systemic corticosteroids are utilized if inhaled corticosteroids are not sufficient to relieve inflammation & bronchospasms. Systemic corticosteroids have much higher incidence & severity of adverse effects. Abrupt withdrawal can cause severe to fatal effects. Systemic corticosteroids are also utilized for shock & other non-pulmonary indications with different dosage guidelines.

- **Contraindications for Corticosteroids:** Not for acute episode of bronchospasms. Not for patients with systemic fungal infections.

- **Adverse Reactions to Corticosteroids:** hoarseness, sore mouth &/or throat, thrush (candidiasis), bronchospasm, cough, nausea, fluid retention, weight gain, headache, dizziness, seizures, insomnia, cushingoid state, muscle wasting, & arrhythmias.

Table I. NEUROMUSCULAR BLOCKING AGENTS

Generic Name Brand Names	Normal Adult Dosage	Duration of Action	Onset of Action
pancuronium Pavulon **Nondepolarizing	IV 0.4 to 0.1 mg/kg	40 - 60 min	1 min
succinylcholine chloride Anectine *Depolarizing	IV 0.6 to 1 mg/kg	4 - 10 min	< 1 min
tubocurarine chloride Tubarine ** Nondepolarizing	IV 0.165 mg/kg	30 - 90 min	1 min
vecuronium bromide Norcuron ** Nondepolarizing	IV 0.09 mg/kg bolus 0.01 – 0.015 mg/kg maintenance	15 - 25 min	1 min

Table I. Notes: Neuromuscular blocking agents are utilized to induce skeletal muscle relaxation & paralysis. This facilitates intubation & is an adjunct to

mechanical ventilation on modes such as HFOV & others when the pt must be sedated & paralyzed.
* Depolarizing drugs (like Anectine) have no antidote for reversal.
**Nondepolarizing drugs have the antidotes: Neostigmine or Tensilon.

Table J. AEROSOLIZED ANTI-INFECTIVE MEDICATIONS

Generic Name Brand Names	Normal Adult Dosage	Notes
pentamidine isethionate NebuPent Pentam 300	300 mg in 6 cc sterile water Q4 weeks via Neb	Antiprotozoal drug for prophylaxis & treatment of PCP. Give B2 bronchodilator (i.e. albuterol) first to prevent bronchospasms. Pt should use filtered neb like Respirgard II or isolation booth/room. Clinician should use protective barrier mask.
ribavirin Virazole	6 g in 300 cc sterile water via continuous neb for 12-18 Hr/day for 3 to 7 days	Antiviral utilized for prophylaxis & treatment of RSV, & for patients at risk for severe infection. Administer ribavirin with large-volume Small Particle Aerosol Generator (SPAG), or other approved device.
tobramycin Nebcin TOBI	Nebulizer 80 mg BID for 28 days	Antibiotic utilized to treat CF & many bacterial pulmonary infections. Give B2 bronchodilator (i.e. albuterol) first to prevent bronchospasms. May need to increase flowrate to 8-12 lpm depending on solution viscosity.

Table K. OTHER RESPIRATORY & ADJUNCT MEDICATIONS

CARDIAC DRUGS

Generic Name Brand Names	Indications Effects
adrenaline atropine epinephrine	Indications: bradycardia. Effects: increases heart rate.
adenosine Adenocard propranolol Inderal verapamil Calan	Indications: tachycardia. Effects: decreases heart rate.
digoxin Lanoxin	Indications: atrial fib/flutter, paroxysmal supraventricular tach (SVT), & CHF. Effects: antiarrythmic, decreases heart rate, strengthens the force of cardiac contractions.

51

isosorbide dinitrate
  Isordil
nitroglycerine
propranolol
  Inderal
	Indications: angina.
	Effects: relieves chest pain.

Cardiac drugs have other indications & effects in addition to those listed in this table.

DIURETICS

Generic Name (Brand Names)	Indications Effects
acetazolamide (Diamox) bumetanide (Bumex) diuril (Chlorothiazide) furosemide (Lasix)	Indications: pulmonary edema, generalized edema, hypertension, hypervolemia. Effects: Increases urine output. Can relieve dyspnea related to fluid retention.

NITRIC OXIDE GAS (NO)
   See Chapter 5, Section A: Medical Gases

SEDATIVES & ANALGESICS

Generic Name (Brand Names)	Notes

Sedatives - Tranquilizers
   Benzodiazepines
   diazepam (Valium)
   lorazepam (Ativan)
   chlordiazepoxide (Librium)
   propofol (Diprivan)
	Utilized to sedate & relieve anxiety

Narcotic Analgesics - Opiates
   codeine
   fentanyl (Sublimaze)
   meperidine (Demerol)
   propoxyphene (Darvon)
   morphine (Duramorph, MS Contin)
	Utilized to sedate & relieve pain. Morphine can also increase exercise tolerance in COPD.

Analgesics relieve pain. Sedatives generally: produce sedation, anesthesia, muscle relaxation, anxiety relief, &/or sleep. Sedatives cause CNS depression & can cause respiratory depression. Clinicians must monitor vital signs & respiratory rate.

SURFACTANT

Generic Name (Brand Names)	Notes
beractant (Survanta) calfactant (Infasurf) poractant alfa (Curosurf) *lusupultide (Venticute) is a surfactant pending FDA approval for neonates, & for adults with pneumonia or ARDS.*	Surfactant is indicated for prevention & treatment of IRDS in premature infants. Dose varies by manufacturer, & is based on body weight at birth. Surfactant is given intratracheally to each lung zone. Surfactant for treatment of adult ARDS, pneumonia, & other diseases are currently undergoing clinical trials.

VASODILATORS

Generic Name (Brand Names)	Indications Effects
atenolol (Tenormin) captopril (Capoten) clonidine (Catapres) nitroprusside (Nipride) prazosin (Minipress)	Indication: hypertension. Effects: decreases blood pressure.

VASOPRESSORS

Generic Name (Brand Names)	Indications Effects
dobutamine (Dobutrex) dopamine (Intropin) epinephrine norepinephrine (Levophed)	Indication: hypotension. Effects: increases blood pressure.

Section A: O$_2$ & OTHER MEDICAL GASES
**Oxygen Therapy: O$_2$ Delivery Devices, Tank factors,
Oxygen Index (OI) & P/F Ratio,
Helium, HBO, Nitric Oxide**

Section B: HUMIDITY & AEROSOL THERAPY

Section C: MEDICATION DELIVERY MODES
Nebulizer, IPPB, MDI, DPI, Spacer

Section D: CPT, PEP, IS & other LUNG HYGIENE THERAPY
CPT, PD, PEP, PAP, Flutter, Vibratory Vest, IS

O$_2$ & OTHER MEDICAL GASES
Oxygen Therapy: O$_2$ Delivery Devices, Tank factors, Oxygen Index (OI) & P/F Ratio, Helium, HBO, Nitric Oxide

OXYGEN THERAPY

Oxygen Delivery Device	Oxygen Flow Rate LPM	FiO$_2$ % (appx)	Notes
NC			
Nasal Cannula (NC)	1 - 6	24% to 44%	NC is a low flow system. 1 lpm=24%, 2 lpm= 28%, 3 lpm= 32%, 4 lpm= 36%, 5 lpm= 40%, 6 lpm= 44%
Oxymizer (a type of oxygen conserving NC)	1 – 6 (or higher PRN)	28% to 46%	Delivered FiO$_2$ is higher than regular NC due to reservoir; meaning a lower O$_2$ flow can be utilized w/same benefit to pt. In the clinical setting, O2 flow > 6 is utilized PRN. Humidifier is built in; don't add humidity, as it can damage the device.
High Flow Heated NC Adult............. Neonate........	....5-40* 1-8*	21% to 100%	Specially designed nasal cannula or trans-tracheal cannula used to provide heated humidity. For adults, can use at flowrates up to 40L/m* nasal & 20L/m* tracheal.
MASK & Misc			
Simple Mask	5 - 8	40% to 60%	Minimum flow of 5 lpm needed to flush CO$_2$ from mask
Venturi Mask	Varies*	24% to 50%	Set O$_2$ flowrate per guidelines to attain precise FiO$_2$.
Aerosol Mask w/Venturi	Varies*	21% to 100%	Set O$_2$ flowrate per guidelines to attain precise FiO$_2$. Includes face mask, face tent, or trach collar, either heated or cooled.
Partial Rebreather Mask (PRB)	8 - 15	60% to 80%	PRB is a nonrebreather with flaps removed. Keep reservoir bag at least 2/3 full using O$_2$ flow of at least 8 lpm to flush out CO$_2$.
Nonrebreather Mask (NRB)	8 – 15 or flush	90-100%	Highest FiO$_2$ achieved by using 3 flaps, but for pt safety, 1 flap is usually left off. Keep reservoir bag

			at least 2/3 full using O_2 flow of at least 8 lpm to flush CO_2 out.
Resuscitation (Ambu) Bag	15 – flush	100%	Delivers ventilation & O2 to pt. Good for emergency & transport.
Oxygen Conserving Device (OCD)	1 - 6	24% – 44%	O_2 delivered on inspiration only. FiO_2 & SpO_2 may be less when using device, so monitor closely for dyspnea. Normally used w/a only.
Oxygen Concentrator**	1 - 6	**	Uses a molecular sieve to produce O_2 from room air. Used in home care/alternate care when wall O_2 unavailable. Delivers FiO_2 slightly less than NC. See note below.**

*Flowrate of O_2 required to attain precise FiO_2 varies by manufacturer. Follow guidelines set forth by manufacturer as well as employers policy when setting flowrate.

**Concentrator FiO_2 output can range from 92% to 100% (dependent on age & type of equipment), whereas high-pressure wall outlet oxygen source delivers FiO_2 at 100%. Concentrator can also be used with mask O_2, however, one may need to connect two concentrators to deliver adequate flowrate.

Note: High flow devices meet total patient demand for minute volume. High flow devices include: Venturi mask, Aerosol mask w/venturi, & Ambu bag. NRB is sometimes considered high flow if used at high flow rate with flaps intact.

Oxygen tank factors (L/psi) :

E cylinder **.28** (for quick estimate, round to .3)
G cylinder **2.4**
H cylinder **3.1** (for quick estimate, round to 3)

Oxygen tank duration of flow calculations (in minutes) :

(gauge pressure in psi X   tank factor) / liter flow in lpm

Example: How long will a full E cylinder at 2200 psi last if running at 5 lpm?
(2200 X .28) / 5 =   123 minutes or 2 hour, 3 minutes

Oxygen & Air Entrainment Ratios:

This table describes the number of parts of oxygen, and the number of parts of air, that are entrained at different O_2 percentages. Note that 60% is a 1:1 ratio.

FiO_2	Entrainment Ratio Oxygen : Air		FiO_2	Entrainment Ratio Oxygen : Air
24%	1 : 25		60%	1 : 1
28%	1 : 10		70%	1 : 0.6
35%	1 : 4.3		100%	1 : 0
40%	1 : 3			

Oxygen Index (OI)

OI is a clinical tool utilized for ventilator patients to determine the degree of hypoxemia, to monitor for improvement, and as a weaning tool. OI takes into account the mean airway pressure (Mean Paw), FIO_2, and PaO_2. It is common knowledge that it's not good to be on a high FIO_2 and have low PaO_2, & the oxygen index formula helps by placing a number on the severity of the hypoxemia.

Formula: Oxygen Index = (Mean P_{aw} x FIO_2) / PaO_2

Sample) Given: Mean P_{aw} 15, FIO_2 75%, PaO_2 = 50 mmHg

Oxygen Index = (15 x 75) / 50 = 23

Evaluating the Oxygen Index (OI) Results:

OI	Significance
5 or less	Good Oxygenation
6-19	Mild to Moderate Hypoxemia
20 or more	Severe Hypoxemia

PaO$_2$/FiO$_2$ Ratio   (P/F Ratio)

The P/F ratio is a tool utilized to assess the degree of hypoxemia for intubated and non-intubated patients. The only factors are the PaO$_2$ and FIO$_2$. Note: FIO$_2$ in decimal form is utilized for this formula.

Formula: P/F Ratio = PaO$_2$ / FIO$_2$

   Sample: Find the P/F ratio of a pt on 40% VM and a PaO$_2$ of 70.
            70 / .40 = 175

Evaluating the P/F Ratio Results:

P/F	Significance
Above 200	Normal Oxygenation
100-200	Moderate Hypoxemia
Under 100	Severe Hypoxemia

HELIUM THERAPY

Helium is a low-density gas utilized to ease WOB in cases of large airway obstruction. This includes asthma, COPD, croup, & stridor. A helium/oxygen mixture is normally given via nonrebreather or simple mask. A common short-term side effect is voice distortion. Two mixtures available of helium/oxygen are:
   80% helium / 20% oxygen or
   70% helium / 30% oxygen

If using an oxygen flowmeter, the flowrate must be adjusted for helium as follows:
   80/20   Divide desired flowrate by 1.8 for corrected flowrate.
   70/30   Divide desired flowrate by 1.6 for corrected flowrate.
Sample: You want a flowrate of 10 lpm using 70/30 helium/oxygen gas. What is the flowrate to set on the oxygen flowmeter?
         10 / 1.6 = 6.25 lpm

HYPERBARIC OXYGEN THERAPY (HBO or HBOT)

HBO is oxygen therapy at pressures greater than one atmosphere. Atmospheric pressure absolute (ATA) is a unit of measure utilized in HBO.   **1 ATA = 760 mmHg**
HBO therapy is often utilized at 2 to 6 ATA. The duration of treatments & frequency vary greatly based on condition.

HBO using 100% oxygen increases oxygenation (PaO_2) <u>much more</u> than 100% oxygen alone. Tissue oxygenation is also greatly increased.

The patient breathes 100% oxygen during HBO therapy. Pressurization is achieved using an HBO chamber, either monoplace or multiplace. The beneficial aspect of the multiplace chamber is that HBO professional staff can accompany the pt and provide care during treatment.

HBO is indicated to treat:
Air or gas embolism, decompression sickness, carbon monoxide* or cyanide poisoning, gangrene, necrotizing soft tissue infection, ischemic skin grafts, radiation necrosis, & extreme blood loss.
HBO is also used to expedite wound healing & promote neovascularization (due to HBO's hyperoxia effects).
Also, to treat cerebral edema, crush injuries, & burns (via HBO's effects of hyperoxia & vasoconstriction that decrease edema while increasing tissue oxygenation).

*HBO is an expedient method of removing CO (as treatment for smoke inhalation) from the blood; about 22 minutes using 100% O_2 at 3 ATA. Utilizing 100% oxygen without HBO takes much longer.

Hazards & complications of HBO include:
Ear or sinus trauma/ tympanic membrane rupture (it's important to give instructions regarding "clearing the ears" to prevent this.
Pneumothorax, air embolism, oxygen toxicity, CNS toxicity, sudden decompression, claustrophobia, & temporary visual changes.
Also, the risk of fire. Fire prevention in the chamber is important.
Minimizing risk of fires includes: **Do not use** alcohol or petroleum products, sprays, makeup, or deodorant. Also, one should wear 100% cotton clothing to avoid static electricity.

NITRIC OXIDE (NO)

Nitric Oxide gas is a pulmonary vasodilator that effectively increases oxygenation. It is utilized to treat pulmonary hypertension & hypoxic respiratory failure. The current therapeutic range for delivering NO is 2 to 30 ppm. (Higher levels up to 80 ppm are undergoing clinical trials and have shown effectiveness in certain cases. However, at the higher levels, one must closely monitor for adverse effects).

Nitric Oxide effectively increases oxygenation in cases of severe hypoxemia including:
ARDS, COPD, lung transplantation (& other organ transplants), & neonatal applications to name a few.

Adverse Reactions to Nitric Oxide include:
Increased methemoglobin (MetHb);
     the goal is to maintain MetHb well below 3%.
Increased nitrogen dioxide (NO_2), a toxic byproduct of NO therapy;
     the goal is to maintain NO_2 below 1 ppm.
Pulmonary edema, platelet inhibition, cellular damage, and rebound problems.

Note: higher levels of NO parallel with higher risk of adverse effects.
Note: NO is not the same as nitrous oxide, which is N_2O (laughing gas).

Notes:

HUMIDITY & AEROSOL THERAPY

AH or Absolute Humidity
The actual content of water present in a given volume of gas, expressed in mg/L or as water vapor pressure (PH_2O).

**The maximum AH at 37° C is 44 mg/L
& PH_2O is 47 mm Hg at 100% RH**

AH / 44 mg/L x 100 = % Body Humidity

RH or Relative Humidity
The comparison of the actual amount of water in a volume of gas (AH) to the maximum capacity of water in the same volume of gas, at a given temperature. When the humidity content equals the capacity, the gas is 100% saturated, meaning RH is 100%.
% RH = AH / Capacity X 100
Humidity deficit = Capacity – AH

Capacity or Maximum Absolute Humidity or Potential Humidity
The maximum amount of water a volume of gas can hold at a given temperature. Capacity is temperature dependent. Utilizing higher temperatures, one can increase humidity & PH_2O.

Alveolar water vapor content at 37° C (body temperature) is 44 mg/L under ideal conditions, with PH_2O 47 mm Hg (100% RH). Our goal is to deliver 80-100% body humidity.

Temperatures & Water Vapor Content

Temperature °F ... °C	Water Vapor Content (mg/L)	Partial Pressure (mm Hg)
32 ... 0	4.9	4.6
50 ... 10	9.3	9.1
70 ... 21	18.0	18.7
98.6 37	44.0	47.0

Adequate humidity is important to the pulmonary system. Insufficient humidification can result in: Dry, thick, & retained secretions, decreased ciliary activity, atelectasis, & pneumonia.

If the room temperature is about 75°F, the water vapor content is about 20 mg/L (see table above). This represents a significant humidity deficit of about 24 mg/L when compared to alveolar water vapor content at 37°C. This is acceptable for a low flow nasal cannula, since the nose acts as a heat & humidity source. However this represents a significant water vapor deficit if delivered directly through an ET or Trach tube. One must add a heated humidifier to achieve 80-100% body humidity.

Aerosol Particle Deposition
An aerosol is a visible spray of particulate matter suspended in a gas. In contrast, humidity is an invisible vapor. The optimal size for aerosol particle deposition in the alveoli is 0.5 to 3 microns, with the therapeutic size ranging from 0.5 to 10 microns. The larger particles deposit higher in the respiratory tract.

Notes:

MEDICATION DELIVERY MODES
Nebulizer, IPPB, MDI, DPI, Spacer

Medication Delivery Device	Average Duration of Treatment	Oxygen Flow Rate LPM*	Notes
Aero Eclipse	3 - 5 min	8	Activated by inspiratory breath. Can use cont flow PRN. Use concentrated med solution.
DPI Dry Powder Inhaler	5 min	...	Device to deliver medication in dry powder form. Instruct pt to hold breath for 5-10 seconds after inhalation. If more than 1 puff, wait at least 1 min between puffs.
IPPB** Intermittent Positive Pressure Breathing	10-15 min	...	Lung expansion therapy to treat or prevent atelectasis, improve ventilation, & improve lung function. Usually given w/med, but can use saline. 50 psi source gas needed; oxygen or air. Adjust flowrate & PIP (avg range 15-24 cmH$_2$O) based on pt needs & tolerance.
MDI Metered Dose Inhaler	5 min	...	Delivery of a metered (measured) dose of med for inhalation. If it is a liquid med using a propellant aerosol for delivery, one must shake the container well before each puff. Then inhale 1 puff at a time & hold breath 5-10 seconds. Wait at least 1 min between puffs. Also see DPI; dry powder meds don't need to be shaken.
NEB or SVN Nebulizer, Small Volume reservoir	10 min	6 - 8	This is the typical nebulizer treatment. Optional delivery methods are hand-held, mask, or in-line w/ vent, bipap, or cpap.
Nebulizer, Continuous Large reservoir Heart............... MiniHeart.........	.....7-8 hr .....3-4 hr	...10..... .....2.....	Continuous neb treatment. Med & saline amounts vary based on MD order for med mg/hr, & manufacturer instructions for amount of saline & flow rate. MiniHeart can be given in-line w/vent, cpap, bipap, or aerosol.

Spacer	...	...	Used with MDI's, and <u>certain</u> DPI's. Check manufacturer instructions for compatibility. Improves medication deposition in the lungs & lessens deposition in the mouth.

* Alternately use an air compressor with preset flow rate of 6-8 lpm when applicable.
** IPPB contraindicated in untreated pneumothorax.

Notes:

CPT, PD, PEP, PAP, Flutter, Vibratory Vest, IS
LUNG HYGIENE THERAPY

Therapeutic Modality**	Duration & Frequency of Treatment (avg)	Notes
CPT / PD	Duration: 10-15 minutes. Frequency: QID	Chest Physical Therapy (CPT) is percussion & vibration to problematic areas of the lungs. Postural Drainage (PD) is gravity drainage of target areas of the lungs. CPT/PD is utilized to prevent & reverse atelectasis, improve mobilization & removal of secretions, & improve lung function. Encourage cough after tx. (See CPT/PD positions next page).
PEP Acapella or Flutter	Duration: 5-20 minutes. Frequency: QID to Q2 hr w/a	PEP is an alternate method for CPT. PEP produces vibratory/oscillatory positive expiratory pressure therapy. Utilized to treat or prevent atelectasis, mobilize & remove secretions, lessen air trapping in COPD, & improve lung function. The pt blows into device. Oscillation rate is adjustable: on Acapella by rotating knob, on Flutter by angle of tilt. Neb can be given in-line w/Acapella. Encourage cough after tx.
PAP EZ PAP	Duration: 10 minutes. Frequency: QID	PAP is positive airway pressure lung expansion therapy to treat or prevent atelectasis & improve lung function. Utilized w/ mask or mouthpiece. Oxygen or air flow starting at 5 lpm & adjust up to max 15 lpm based on pt needs and tolerance. Manometer in-line used to measure pressure (avg 10-20). Can give alone w/source gas, or add neb in-line. Encourage cough after tx.
Vibratory Vest	Duration: 30 minutes. Frequency: QID	The vibratory vest is an alternate for CPT that provides chest wall oscillations. Utilized to improve lung function & airway clearance by mobilizing & facilitating removal of retained secretions in diseases like CF & COPD. To begin tx: Select appropriate vest size. Set pressure to 4-6. Set frequency to 10,12, or 14 Hz, starting at low

		frequency, & adjusting upward Q 10 minutes based on pt need & tolerance. Encourage cough after treatment.
IS Incentive Spirometry	10 breaths Q1-2 hr w/a	Description: Lung expansion therapy to prevent & treat atelectasis. The normal value is 50 mL/kg.

**Therapy Precautions & Possible Contraindications:
Clinically unstable, unstable cardiac or hemodynamic status, increased ICP, recent neurosurgery, pulmonary emboli, empyema, rib fracture, pneumothorax, hemoptysis, conditions prone to hemorrhage, & recent postop or injury w/severe pain.
Contraindications for CPT & Vest: Untreated pneumothorax, rib fracture, pulmonary empyema & severe pain. Also see Respiratory Medications chapter for indications & contraindications of specific medications.

CPT/PD Positions:

Upper Lobes, Apical Segments
   Position: Patient reclines back on pillow at 30° angle.
   Percuss: Between clavicle & top of scapula on both sides.

Upper Lobes, Anterior Segments
   Position: The bed is flat. Patient lies flat on the back.
   Percuss: Between clavicle & nipple on both sides.

Upper Lobes, Posterior Segments
   Position: Patient sits & leans forward on pillow at 30° angle.
   Percuss: Over the upper back on both sides.

Right Middle Lobe, Medial/Lateral Segment
   Position: The foot of the bed is elevated 15°. Patient lies head down, on left side, rotated ¼ turn backward.
   Percuss: Under the breast on right side.

Left Upper Lobe, Lingular Segment
   Position: The foot of the bed is elevated 15°. Patient lies head down, on right side, rotated ¼ turn backward.
   Percuss: Between nipple & armpit on left side.

Lower Lobes, Anterior Basal Segments
   Position: The foot of the bed is elevated 30°. Patient lies head down, on right side, then on left side.
   Percuss: Over lower ribs just below the axilla on both sides.

Lower Lobes, Lateral Basal Segments

Position: The foot of the bed is elevated 30°. Patient lies head down, on right side, then on left side, rotated ¼ turn inward toward bed.
Percuss: Over lower ribs on both sides.

Lower Lobes, Posterior Basal Segments

Position: The foot of the bed is elevated 30°. Patient lies head down, on stomach, with pillow under hips.
Percuss: Over lower ribs, close to the spine, on both sides.

Lower Lobes, Superior Segments

Position: The bed is flat. Patient lies prone on stomach with 2 pillows under hips.
Percuss: Over the middle of the back, at the tip of the scapula, on both sides of the spine.

Notes:

Chapter 6

DIAGNOSTIC TOOLS

Pulse Oximetry
Transcutaneous Oxygen (TcPO$_2$)
Capnography (ETCO$_2$)

Pulse Oximetry

Pulse oximetry is a reliable non-invasive technique for continuous or quick measurement of oxygen saturation (SpO$_2$). Under ideal conditions, the trend in SpO$_2$ closely parallels ABG SaO$_2$. The normal SpO$_2$ is 95% - 100%, but varies with age (see ABG chapter). Pulse oximetry utilizes a spectrophotometer to measure percentage of oxyhemoglobin. The pulse rate is also displayed. Some options for placement of the pulse oximeter include the ear, finger, or toe.

Pulse oximetry is most accurate in patients with good perfusion, who have actual saturation of 70% or higher. An accurate reading is indicated by good waveform, and a correlation between the palpable pulse & pulse rate displayed on oximeter.

The following is a list of factors that can give erroneous saturation readings, along with possible remedies in parentheses:
- poor perfusion (correct underlying condition)
- hypothermia (correct underlying condition, warm the probe site)
- motion artifact (keep probe site still if possible, or use a different site)
- artificial fingernails (use a different site)
- nail polish (remove nail polish or use a different site)
- abnormal hemoglobin like methemoglobin or carboxyhemoglobin (correct underlying condition)
- loose probe (apply tape or change probe if faulty)
- an unclean site or an unclean probe (clean site &/or probe with alcohol & dry with gauze, then replace probe)

Transcutaneous Oxygen Monitoring (TcPO$_2$)

TcPO$_2$ is a measure of the partial pressure of oxygen on the skin. The TcPO$_2$ device utilizes a heated anode that improves local vasodilation; thus allowing outward diffusion of O$_2$, arterialization of skin blood vessels, and the PO$_2$ reading. The electrode measuring TcPO$_2$ is polargraphic & works like the Clark electrode. The device must be maintained, cleaned, & calibrated, on a regular schedule.

The normal $TcPO_2$ is the same normal value as PaO_2 at 80-100 torr for adults. Depending on equipment, there may be a delayed response time of 5-15 seconds during continuous monitoring, and a longer delay time at initial setup.

The site chosen should have good capillary circulation and no large vessels in the area. For neonates, the best areas include: thoracic area, sternal area, stomach, thighs, or buttocks. For adults, the inside of an arm or subclavicular region can be used.

The site should be free of hair and cleaned with alcohol. Depending on the type of electrode, a small drop of water, gel, or solution is applied to the electrode to form a good seal with the skin. The electrode should be moved periodically to prevent burns.

$TcPO_2$ has proven most beneficial in neonates. In adults, thicker skin & skin texture adversely affects accuracy of readings. $TcPO_2$ has also proven beneficial in HBO therapy (even for adults). This is due to the ability of HBO to facilitate very high increases in PaO_2. $TcPO_2$ is used in HBO to trend large variances in PO_2 (i.e. a change from 80 to 400). During HOB, a regular site as listed above can be used to get an overall reading of blood PO_2, and/or a specific site of injury can be monitored for desirable increases in PO_2.

Capnography (ETCO$_2$)

Capnography is the measurement of exhaled carbon dioxide or end-tidal CO_2 (ETCO$_2$). This measurement can be expressed as a percentage of CO_2 or in mmHg. The normal percentage is 5% to 6% CO_2, which equals 35-45 mmHg. CO_2 reflects cardiac output & pulmonary blood flow. The CO_2 is transported by the venous system to the right side of the heart & then to the lungs. When CO_2 diffuses out of the lungs with the exhaled air, the capnometer measures alveolar end-tidal CO_2.

Small & portable ETCO$_2$ detectors can be used immediately post intubation to help detect if the ET tube is in the lungs. The ETCO$_2$ detector attaches to the ET tube. If the percentage displayed is in the 5% to 6% range (usually detected by color change on the device), then one can be fairly certain the tube is in the lungs. However, one should follow all the steps in assessing for successful intubation as outlined in the intubation section of this book.

If the reading is less than normal (or no color change on the device), clinical skills must be called upon to assess the situation. Sometimes a false negative can occur. (For instance, in the case of a patient in cardiopulmonary arrest for an extended period of time.) Please review the intubation section of this book for other ways to assess for successful intubation.

Capnography monitors are also available for continuous CO_2 monitoring. These devices have digital displays & alarms. An infrared light beam measures CO_2. Capnometers are often utilized in-line with mechanical ventilation. These devices can give an estimation of $PaCO_2$ and are very useful to _trend_ the changes in $PaCO_2$. ABG's can be utilized for a baseline correlation, and also for periodic comparison. If the $ETCO_2$ changes, the blood gas $PaCO_2$ is changing, and further assessment/action is needed.

Important notes: Capnometers must be maintained, cleaned, & calibrated, on a regular schedule. Normal $ETCO_2$ is zero during inspiration. Mechanical deadspace is probably present if the reading does not return to zero during inspiration. $ETCO_2$ can drop to zero during apnea, complete airway obstruction, ventilator malfunction, or capnograph malfunction.

NOTES:

Chapter 7

ABG's & Formulas

7-a > Blood Gas Formulas
7-b > Oxygen Index (OI)
7-c > PaO₂/FiO₂ Ratio (P/F Ratio)
7-d > Relationship of pH, PCO₂, and HCO₃
7-e > Normal ABG's & VBG's
7-f > Estimating Normal PaO₂ based on Pt Age
7-g > Estimating Saturation for a given PO₂
7-h > Interpretation of Arterial Blood Gases
7-i > Most Common Causes of Abnormal ABG's
7-j > COPD Patient Normal Values
7-k > O₂ – Hb Dissociation Curve
7-l > ABG Sampling Sites
7-m > Sampling Errors
7-n > ABG Machine Electrodes

ALSO SEE: Hemodynamics w/Formulas – Chapter 8
Mechanical Ventilation Formulas – Chapter 16
Misc. Formulas & Conversions – Chapter 20

7-a > BLOOD GAS FORMULAS

Parameter & Abbrev.		Normal Value	Calculations & Notes
A-a gradient or Alveolar to arterial O_2 difference	$P(A\text{-}a)O_2$ or $A\text{-}aDO_2$	< 20 torr on room air..... 25-65 torr on O_2 100%	$PAO_2 - PaO_2$ Evaluate for presence of diffusion defect, % shunt, & effectiveness of oxygen therapy.
Alveolar air equation or Alveolar PO_2	PAO_2	100 torr on room air	$(PB\text{-}PH_2O) \times FiO_2 - (PaCO_2 / 0.8)$ Sample: $(760\text{-}47) \times .21 - (40 / 0.8) = 100$
Arterial to venous O_2 content difference	$C(a\text{-}v)O_2$	3.5-5 vol%	$CaO_2 - CvO_2$ Index of tissue oxygenation.
Arterial to venous O_2 pressure difference	$Pa\text{-}vDO_2$	50 mmHg	$PaO_2 - PvO_2$ Index of tissue oxygenation.
Carbon dioxide production	VCO_2	200 mL/Min	Affected by many bodily factors like metabolism, substrate, and buffering.
CO_2 Exhaled; partial pressure of carbon dioxide in expired gas	$PECO_2$	30 mmHg	Measured by collecting expired gas in large bag & using an infrared CO_2 analyzer to measure PCO_2. Normally slightly less than $PaCO_2$. ↑ $PECO_2$ = ↓ ventilation or ↑VD/VT.
Content of oxygen in arterial blood	CaO_2	15-20 vol%	$(Hb \times 1.34 \times SaO_2) + (PaO_2 \times .003)$

Content of oxygen in mixed venous blood	CvO_2	12-15 vol%	$(Hb \times 1.34 \times SvO_2) + (PvO_2 \times .003)$
Deadspace to tidal volume ratio	VD/VT	20-40% if off ventilator -------- 40-55% if on ventilator	$(PaCO_2 - PECO_2) / PaCO_2$ Measures deadspace ventilation or ventilation w/o perfusion. ↑ in pulmonary embolism, pulm tumor, pulm HTN, & rapid shallow breathing. VD Anatomical = 1 mL/pound. VD Mechanical = 10 mL/inch.
Oxygen consumption	VO_2	250 mL/Min	$VO_2 = CO \times C(a-v)O_2$ Metabolic rate or amount of O_2 used by one's body.
Oxygen transport	---	1000 mL/Min	Used to assess amount of oxygen delivered to tissues.
Respiratory Quotient	RQ	0.7 – 1.0	VCO_2 / VO_2 . Ratio of CO_2 production to O_2 consumption.
Shunt %	QS/QT	Normal 5% -------- Moderate Defect 10-20% -------- Severe defect > 20%	$(A-aDO_2) \times .003 / (CaO_2 - CvO_2) + (A-aDO_2) \times .003$ Cardiac output that is shunted, or perfusion w/o ventilation. Refractory hypoxemia. ↑ in atelectasis, pneumonia, pulmonary edema, & anatomical/heart defects.

Note: All parameters are adult values unless noted.

7-b > OXYGEN INDEX (OI)

OI is a clinical tool utilized for ventilator patients to determine the degree of hypoxemia, to monitor for improvement, and as a weaning tool. OI takes into account the mean airway pressure (Mean Paw), FIO_2, and PaO_2. It is common knowledge that it's not good to be on a high FIO_2 and have low PaO_2, & the oxygen index formula helps by placing a number on the severity of the hypoxemia.

Formula: Oxygen Index = (Mean P_{aw} x FIO_2) / PaO_2

Sample) Given: Mean P_{aw} 15, FIO_2 75%, PaO_2 = 50 mmHg

Oxygen Index = $(15 \times 75) / 50 = 23$

Evaluating the Oxygen Index (OI) Results:
  OI   **Significance**
 5 or less ....Good Oxygenation
 6-19 ....Mild to Moderate Hypoxemia
 20 or more ....Severe Hypoxemia

7-c >   PaO$_2$/FiO$_2$ Ratio   (P/F Ratio)

The P/F ratio is a tool utilized to assess the degree of hypoxemia for intubated and non-intubated patients. The only factors are the PaO$_2$ and FIO$_2$. Note: FIO$_2$ in decimal form is utilized for this formula.

Formula: **P/F Ratio =   PaO$_2$ / FIO$_2$**

Sample: Find the P/F ratio of a pt on 40% VM and a PaO$_2$ of 70.
70 / .40 = 175

Evaluating the P/F Ratio Results:

P/F	Significance
Above 200	Normal Oxygenation
100-200	Moderate Hypoxemia
Under 100	Severe Hypoxemia

7-d >   Relationship of pH, PCO$_2$, and HCO$_3$
in Acute Conditions

pH	PaCO$_2$ (mmHg)	HCO$_3$ (mEq/L)
↓ 0.05	↑ 10	↑ 1
↑ 0.10	↓ 10	↓ 2

Bicarbonate Administration to Correct pH:
One ampule of bicarbonate (44 mEq/L) will result in approximately 0.1 increase in pH.

7-e >   ABG & VBG Adult Normal Values

ARTERIAL BLOOD GASES

Parameter	Normal Value	Normal Range
pH	7.40	7.35 – 7.45
PaCO$_2$	40 mmHg	35 – 45 mmHg

73

PaO$_2$	100 mmHg	80 – 100 mmHg
HCO$_3$	24 mEq/L	22 – 26 mEq/L
BE	0	+ or - 2
Hb	14 g/dL	12 – 18 g/dL
O$_2$ content	20 vol%	15-20 vol%
SaO$_2$	98%	> 95%
COHb	0	< 2%
MetHb	0	<2%

MIXED VENOUS BLOOD GASES

Parameter	Normal Value	Normal Range
pH	7.37	7.32 – 7.42
PvCO$_2$	45 mmHg	40 – 50 mmHg
PvO$_2$	40 mmHg	38 – 42 mmHg
HCO$_3$	24 mEq/L	22 – 26 mEq/L
SvO$_2$	73%	70-75%

7-f >   Estimating Normal PaO$_2$ based on Pt Age

Formula for Estimating Normal PaO$_2$ (useful for age 10 to 80):
    110 – ½ age = Normal PaO$_2$

7-g >   Estimating Saturation for a given PO$_2$

Formula for Estimating Saturation: The 4,5,6,7,8,9 rule:

PO$_2$ mmHg	Saturation %
40	70
50	80
60	90

7-h > Interpretation of Arterial Blood Gases

pH	PaCO$_2$	HCO$_3$	Interpretation
↓	↑	N	Acute Respiratory Acidosis (aka Acute Ventilatory Failure)
N	↑	↑	Chronic or Compensated Respiratory Acidosis (aka Chronic Ventilatory Failure)
↓	↑	↑	Acute superimposed on chronic Respiratory Acidosis/ Ventilatory Failure
↑	↓	N	Acute Respiratory Alkalosis/ Hyperventilation
N	↓	↓	Chronic or Compensated Respiratory Alkalosis/ Hyperventilation
↓	N	↓	Acute Metabolic Acidosis
N	↓	↓	Compensated Metabolic Acidosis
↑	N	↑	Acute Metabolic Alkalosis
N	↑	↑	Compensated Metabolic Alkalosis

N.........Normal
aka......also known as

7-i > Most Common Causes of Abnormal ABG's

Respiratory Acidosis/Ventilatory Failure: Insufficient alveolar ventilation. Pulmonary disease, CNS depression, drugs causing respiratory depression.

Respiratory Alkalosis: Alveolar hyperventilation. Stress, emotional upset, hypoxia, fever, CNS trauma.

Metabolic Acidosis: Lactic acidosis, ketoacidosis (diabetes), renal failure, diarrhea.

Metabolic Alkalosis: Hypokalemia (most common cause), low chloride, diuretics, corticosteroids, vomiting, nasogastric tube.

7-j > COPD Patient Normal Values

Some patients with COPD exhibit chronic CO_2 retention, and breathe on the hypoxic drive. These COPD patients have different "normal" ABG's than other patients. The "normal" values are a state of compensated respiratory acidosis, with below normal PaO_2. Accurate interpretation of these blood gases can be more difficult, especially when the pt has acute respiratory acidosis superimposed on chronic ventilatory failure. (See chart 7-h)

The following formula can be used to <u>estimate</u> the "normal" $PaCO_2$ for a patient with chronic respiratory acidosis:
   (Current HCO_3 – 25) x 2 + 40 = "normal" $PaCO_2$

Example) Blood gases: pH 7.34 $PaCO_2$ 80 HCO_3 41 PaO_2 88
   The pt is 70 years old, & is on 5 l/m nasal cannula. What is the pt's normal $PaCO_2$? How would you correct the problem?
   His normal $PaCO_2$ is about 72. He needs the supplemental oxygen turned down, he does not need assisted ventilation at this point. The clinician should then monitor the pt for improvement.

7-k > O_2 – Hb Dissociation Curve

<u>**Shift to Right**</u>: Indicates decreased oxygen affinity with lower O_2 content for a given PO_2. Causes of Shift to Right: ↓ pH, ↑ PCO2, ↑ temperature, ↑ 2-3 DPG. The P50 is >27.

<u>**Shift to Left:**</u> Indicates increased oxygen affinity with higher O_2 content for a given PO_2. Less O_2 is available to the tissues. Causes of Shift to Left: ↑ pH, ↓ PCO_2, ↓ temperature, ↓ 2-3 DPG, CO poisoning, Fetal Hb. The P50 is <27.

The normal P50 is 27, meaning PO_2 27 mmHg = 50% saturation.

7-l > Arterial Blood Gas Sampling Sites

The most common sites for obtaining blood gases are radial, brachial, & femoral. The best choice is the radial artery due to collateral blood flow. The Allen test or modified Allen test is "positive" if collateral blood

76

flow is present. The Allen test should be performed prior to any radial puncture. The procedure for ABG's includes:

- Double verify patient identification
- Use aseptic technique, wear gloves
- Swab the site with alcohol or other prep solution
- Palpate pulse & puncture at an angle
- Apply pressure to the site for 5 to 10 minutes after puncture.
- Apply bandage to site.

7-m > ABG Sampling Errors

The following are potential blood gas sampling errors that can cause erroneous results:

Sampling Error	Erroneous Results
Air bubbles	pH increases PO_2 increases or decreases toward 150 mmHg (room air value) PCO_2 decreases toward 0 mmHg (room air value)
Sample not iced/ improper cooling	pH decreases PO_2 increases PCO_2 decreases
Too much heparin	pH decreases PO_2 increases or decreases toward 150 mmHg (room air value) PCO_2 decreases toward 0 mmHg (room air value)
Temperature correction not done or incorrect	Pt with increased temp: pH decreased, PO_2 & PCO_2 increased Pt with decreased temp: pH increased, PO_2 & PCO_2 decreased

7-n > Blood Gas Machine Electrodes

PO_2.........Clark electrode
pH...........Sanz electrode
PCO_2......Severinghouse electrode
Other values can be calculated from these results. The co-oximeter gives values for hemoglobin, oxyhemoglobin, carboxyhemoglobin, & methemoglobin. Blood gas machine maintenance & calibration must be performed on a regular schedule for accurate & valid results.

Chapter 8

HEMODYNAMICS

Hemodynamic Parameters, Formulas, Pressures, ICP
Hemodynamic Changes Seen in Diseases
Vital Signs: HR, RR, Temp

ALSO SEE: Related formulas & abbreviations in:
ABG's - Chapter 7
PFT's - Chapter 13
Formulas & Conversions - Chapter 20

HEMODYNAMIC PARAMETERS

Hemodynamic Parameter & Abbreviation		Normal Value (in mmHg unless noted)	Calculations & Notes
Blood Pressure	BP	120/80 Systolic 90-130 Diastolic 60-90	Increased = Hypertension Decreased = Hypotension.
Cardiac Index	CI	3-4.5 L/min/m^2	CI = CO / BSA A measure of cardiac function relative to body size.
Cardiac Output	CO or QT	4-8 L/m	CO = HR x SV / 1000 Used to assess cardiac & hemodynamic function. Also to calculate other parameters. Also see the Fick equation.
Central Venous Pressure	CVP	1-6	Measures pressure in systemic venous system, & mean right atrial pressure. Also estimates right ventricle preload & RVEDP (right ventricular end diastolic pressure). Reflects changes in body fluid levels. CVP is also used to give fluid & drugs.
Ejection Fraction	EF	65-75%	EF = SV / end diastolic volume Cardiac function; % of blood pumped out of the heart per contraction.
Fick equation	Fick	4-8 L/m	Fick equation for cardiac output VO_2 / (CaO_2 - CvO_2 x 10) Used to assess cardiac & hemodynamic function.

Heart Rate	HR	60-100 bpm	Bradycardia < 60 bpm Tachycardia > 100 bpm.
Intracranial Pressure	ICP	Adult 0-15 Child 0-10 Infant 0-5	ICP monitor measures psr of the brain & CSF in intracranial space. ICP ↑ in: CVA, ICH, cerebral edema, traumatic injury to head, brain tumor/mass, hydrocephalus. ↑ ICP can cause ↓ blood flow to the brain, ischemia, & death of brain cells.
Left Atrial Pressure	LAP	5	--
Left Ventricle Pressure	LVP	120/1 Systolic 90-130 Diastolic 1-12	--
Mean Arterial Pressure	MAP	85-100	MAP = systolic BP + (2x diastolic BP) / 3
Mean Pulmonary Artery Pressure	$\overline{PAP}$	10-20	Mean PAP = systolic PAP + (2 x diastolic PAP) / 3
Pulmonary Artery Pressure	PAP	25/8 Systolic 18-30 Diastolic 6-15	Swan-Ganz cath is utilized to measure pressure of blood flow from right ventricle thru pulmonary artery, into lungs.
Pulmonary Capillary Wedge Pressure	PCWP or PWP	4-12	Swan-Ganz cath is utilized to measure back pressure from pulmonary veins, which feed into left side of heart. Estimates left ventricle filling & preload.
Pulmonary Vascular Resistance	PVR	1.5-3 mmHg/L/min or 120-240 dynes	PVR = (Mean PAP – PWP) / CO Multiply result by 80 for dynes. Measures resistance in pulmonary circulation.
Pulse Pressure	--	40	Systolic BP – diastolic BP Measures force of pulse.
Right Atrial Pressure	RAP	2-3	--
Right Ventricle Pressure	RVP	25/1 Systolic 20-30 Diastolic 1-6	--
Stroke Index	SI	40-60 mL/beat/m^2	SI = SV / BSA To assess cardiac function relative to body size.
Stroke Volume	SV	60-130 mL/beat	SV = (CO in lpm / HR) x 1000 Sample 7/80 x 1000=87 mL/beat. To assess cardiac function.
Systemic Vascular Resistance	SVR	15-20 mmHg/L/min or 1000-1600 dynes	SVR = (MAP – CVP) / CO Multiply result by 80 for dynes. Measures resistance in systemic circulation, & gives indication of vascular problems.

Hemodynamic Changes
Seen in Diseases

Disease/ Disorder	CO	CVP	PAP	PCWP	PVR	SVR
ARDS	N	N or ↑	N or ↑	N	N or ↑	N
Cardiogenic shock	↓↓	↑	↑	↑↑	N or ↑	↑
COPD	N or ↓	N or ↑	N or ↑	N	N or ↑	N
Hypervolemia	↑	↑	N or ↑	↑	N	↓
Hypovolemia	↓	N or ↓	N or ↓	↓	N	↑
Left heart failure (CHF)	↓	N or ↑	↑	↑↑	N or ↑	↑
Mitral Valve disorder	↓	N or ↑	↑	↑↑	N or ↑	↑
PEEP too high (causes ↓BP & ↓ venous return)	↓	↓	↑	↑	↑	N or ↑
Pulmonary embolism	N or ↓	↑	↑↑	N	↑↑	N
Right heart failure (Cor pulmonale)	N or ↓	N or ↑	N or ↑	N	N or ↑	N
Septicemia	↑	N or ↓	↓	↓	N	↓

Note: N = Normal

Vital Signs: HR, RR, Temp

Vital Sign	Normal Value	Importance of Values
Heart Rate (HR)	60-100 bpm	Bradycardia < 60 Tachycardia > 100
Respiratory Rate (RR)	8-20/minute	Bradypnea < 8/min Tachypnea > 20/min
Temperature*	Axillary 97.6 °F.......36.4 °C Oral 98.6 °F.......37.0 °C Tympanic 99.6 °F.......37.6 °C Rectal 99.6 °F.......37.6 °C	Hypothermia= low temp Hyperthermia= high temp

* Temperature + or − 1 °F is considered normal.

Chapter 9

LABORATORY TESTS

ALSO SEE: ABG'S – Chapter 7
Hemodynamics – Chapter 8

LABORATORY TESTING - BLOOD

Lab Test	Normal Value (adults)	Potential Diagnoses &/or Complications If ↑ Increased or ↓ Decreased
Albumin	3.5 – 5 g/dL	↓: malnutrition, proteinuria, or leukemia. Note: albumin is a protein component important in nutritional assessment.
Alkaline phosphatase	25 – 125 U/L	↑: liver disease, obstructive jaundice.
Ammonia	12 – 55 µmol/L	↑: liver dysfunction, intestinal disorder.
Amylase	4 – 25 U/mL	↑: pancreatitis.
AST	--	See SGOT
**Bicarbonate, CO_2 content (carbon dioxide), HCO_3	24 – 32 mEq/L	↑: metabolic alkalosis, compensated respiratory failure, CO_2 retention. ↓: metabolic acidosis. Note: This is not ABG $PaCO_2$, this is bicarb.
Bilirubin, total	0.3 – 1 mg/dL	↑: liver disease, hepatitis, biliary obstruction, jaundice.
BUN (Blood Urea Nitrogen) Adult.................. Child.................	8 – 25 mg/dL 5 – 18 mg/dL	↑: kidney failure. ↓: low protein intake, poor nutrition.
**Calcium (Ca^+)	8 – 11 mg/dL	↑: hyperparathyroidism, cancer. ↓: hypoparathyroidism, renal failure, pancreatitis, trauma.
**Chloride (Cl^-)	98 – 106 mEq/L	↑: metabolic acidosis, dehydration. ↓: metabolic alkalosis, diuretics, vomiting, excess n-g suction.
Cholesterol, Total............. LDL.............. HDL	<200 mg/dL <150 mg/dL >40 mg/dL	↑: Total or LDL cholesterol promote premature arteriosclerosis. Note: LDL is low-density lipoprotein, HDL is high-density lipoprotein (known as "good cholesterol").
CO_2 content	--	See Bicarbonate
*CPK (Creatine Phosphokinase).... CPK-MB.................	30 - 150 U/L <4%	↑: myocardial infarction, CVA, trauma, or surgery.

Creatinine	0.5 – 1.3 mg/dL	↑: kidney failure
Digoxin	Therapeutic: 0.8 – 2 ng/mL	Toxic medication level: >2.4 ng/mL
Glucose (FBS- fasting blood sugar)	70-110 mg/dL	↑: hyperglycemia, insufficient insulin, diabetes mellitus. ↓: hypoglycemia due to: inadequate carbohydrate intake, digestive problems, or drug-induced.
Hematocrit	M: 42 - 52% F: 37 - 47%	↑: polycythemia, COPD, dehydration. ↓: anemia, hemorrhage.
Hemoglobin	M: 14 - 18 g/dL F: 12 - 16 g/dL	↑: polycythemia, COPD, dehydration. ↓: anemia, hemorrhage.
HCO_3	--	See Bicarbonate
Ketones	3 mg/dL	↑: diabetic ketoacidosis.
Lactic acid	0.5 – 2 mEq/L	(byproduct of anaerobic metabolism) ↑: hypoxia, hemorrhage, shock, CHF, high protein diet.
*LDH (Lactic Dehydrogenase)	60 – 160 U/L	↑: myocardial infarction, pulmonary embolism, hepatitis.
**Magnesium (Mg^{++})	1.4 – 2 mEq/L	↑: renal insufficiency, tissue trauma. ↓: pancreatitis, malabsorption, alcoholism.
Phosphorus	2.6 – 4.5 mg/dL	↑: renal failure, endocrine disorder. ↓: hyperthyroidism, malabsorption.
Platelets (Thrombocytes)	150,000 – 420,000/mm^3	↑: thrombocytosis, polycythemia vera, granulocytic leukemia, cancer, iron deficiency, spleenectomy, infection. ↓: thrombocytopenia, leukemia, infection, certain cancers, liver disease, low B12 intake, steroids, DIC, chemotherapy.
**Potassium (K^+)	3.5 – 5 mEq/L	↑: metabolic acidosis, kidney failure, excess K+ intake, MI, burns, surgery. EKG may show spiked T wave, cardiac arrhythmias. ↓: metabolic alkalosis, diuretics, diarrhea, vomiting, n-g suction, steroids, malnutrition. EKG may show flat or inverted T wave.
Protein, total	6 – 8.4 g/dL	(important in nutritional assessment) ↓: malnutrition, proteinuria, leukemia.
PT (Prothrombin Time)	10 – 12.5 seconds	↑: abnormal blood clotting factor.
PTT (Partial Thromboplastin Time)	25 – 45 seconds	↑: abnormal blood clotting factor.
RBC (Red Blood Cells)	M: 4.6 - 6.2 F: 4.2 - 5.5 million/mm^3	↑: polycythemia, COPD, dehydration. ↓: anemia, hemorrhage, blood loss.

*SGOT or AST	6 – 40 U/L	Serum Glutamic- Oxaloacetic Trans-aminase or Aspartate aminotransferase. ↑: MI, liver disease, trauma, or surgery.
**Sodium (Na⁺)	135 - 145 mEq/L	↑: high sodium intake, dehydration. ↓: diarrhea, diuretics, renal failure, CHF.
Theophylline	Therapeutic: 10 – 20 mcg/mL	Toxic medication level: >20 mcg/mL.
Triglicerides	35 - 150 mg/dL	↑: increased lipids, conducive to early arteriosclerosis.
*Troponin-T	<0.2 ng/mL	↑: myocardial infarction.
TSH (Thyroid Stimulating Hormone)	0.4 - 5.2 µU/mL	↑: hypothyroidism. ↓: hyperthyroidism.
WBC (White Blood Cells) (Leukocytes)	4,600-10,000/mm³	↑: leukocytosis, leukemia, kidney failure, bacterial infection, MI, injury. ↓: leukopenia, viral infection, aplastic anemia, chemotherapy.

WBC differential

Neutrophils (segs/bands)	50 – 75%	↑: infection, neoplasm, trauma. ↓: autoimmune disease, bone marrow disease, HIV infection.
Lymphocytes	20 – 40%	↑: viral infection, lymphocytic leukemia. ↓: immunodeficiency, AIDS, steroids.
Monocytes	1 – 9%	↑: mononucleosis, monocytic leukemia, viral or fungal infection, TB.
Eosinophils	1 – 5%	↑: asthma, allergies, parasitic infection.
Basophils	0.3 – 2%	↑: leukemia, low thyroid function, CA. ↓: hyperparathyroidism, stress.

Notes:

*Enzymes are substances that stimulate biochemical changes. Certain enzymes are also released in response to injury, & are significantly increased after an MI. Certain enzymes are also increased after damage to: skeletal muscles, liver, brain, kidneys, or other organs. Some important enzymes in the above list include: CPK, LDH, SGOT-AST, & Troponin-T.

**Electrolytes are minerals that become ions in solution & acquire the capacity to conduct electricity. Positively charged ions are cations, & negatively charged ions are anions. Common electrolytes include: bicarbonate, sodium, potassium, chloride, calcium, & magnesium. A complex balance of electrolytes in the body, between intracellular & extracellular fluid, is essential for normal function of cells & organs. Electrolytes regulate hydration & pH, and are essential to nerve &

muscle function. If an electrolyte is at an extreme high or low, severe to fatal complications can result.

Electrolytes are lost through bodily secretions such as sweat, urine, excretion, diarrhea, & vomiting. Proper electrolyte balance is achieved by intake of substances containing electrolytes, adequate kidney function, & normal endocrine system function. The endocrine system includes the pituitary, parathyroid, & adrenal glands. The glands regulate ADH, a hormone that can cause fluid retention as a side effect of positive pressure ventilation.

URINE LABORATORY TESTING

Urine Lab Test	Normal Value	Potential Diagnoses If ↑ Increased or ↓ Decreased
Urine output	1200-2000 mL/day	↑: high fluid intake. ↓: low fluid intake, dehydration, kidney failure. Note: Urine output is dependent on fluid intake; normally intake & output are nearly equal.
Urine pH	4.5 - 8	↑: metabolic alkalosis, low protein diet, infection. ↓: metabolic acidosis, high protein diet.
Urine glucose	Negative	↑: diabetes.
Urine protein	Negative-Trace	↑: kidney disease.
Urine specific gravity	1.005-1.030	↑: concentrated urine. ↓: renal insufficiency.

CSF (CEREBROSPINAL FLUID) LAB VALUES

CSF Lab Test	Normal Value	Potential Diagnoses if Results are Abnormal
CSF color	Clear, odorless	If cloudy: infection, meningitis. If yellow: brain hemorrhage, brain tumor, or spinal cord tumor. If bloody: brain hemorrhage, stroke, or skull fracture.

White cells	0 - 4 per mm^3	↑: infection, meningitis, TB, syphilis.
CSF Pressure (not same as ICP*)	80 – 200 mm H$_2$O	↑: meningitis, edema, hemorrhage, stroke, brain tumor, cyst, infection.

* ICP (intracranial pressure) normal value listed in Hemodynamics chapter.

SPUTUM LAB TESTING

Notes:
Sputum Culture – identifies type of bacteria present.
Sputum Gram Stain – identifies gram positive or gram negative infection.
Sputum Sensitivity – identifies the drug(s) that will destroy the infection.

Sputum Analysis	Potential Diagnoses
Mucoid, thin, clear or white.	Normal if zero to 100 mL/day. Otherwise, possible pulmonary infection, bronchitis.
Yellow, viscous, may be purulent.	Pulmonary infection or disease.
Green, viscous, purulent, bad odor.	Pulmonary infection or disease. Stagnant sputum. Gram negative bacteria.
Red, bloody, blood-tinged.	Hemoptysis. Pulmonary trauma, infection, or disease, bleeding tumor, CF, or TB.
Dark, brown, or rusty red.	Old retained secretions, old dried blood. Pulmonary infection or trauma. Possible healing at source of bleeding, though disease may still be present.
Pink frothy.	Pulmonary edema.

NOTES:

Chapter 10

BRONCHOSCOPY

Description of Bronchoscopy: Direct visual examination of the airways via insertion of a bronchoscope. Useful for diagnosing and/or treating airway problems as listed below.

Types of Bronchoscopy: 1) Flexible Fiberoptic Bronchoscopy
2) Rigid Tube Bronchoscopy

Flexible Fiberoptic Bronchoscopy

Description:
A long thin flexible tube containing fiberoptics that transmits light images as it is advanced into the airways. There are three channels including: the light transmission channel, visualizing channel, & a multipurpose open channel. Flexible fiberoptic bronchoscopy is the most common type used in the hospital setting, outside of the OR. It is performed by a physician with respiratory & nursing assistance.

Advantages over the rigid tube bronchoscope:
Much more comfortable to patients. Can be performed using just local anesthetic. More versatile, and offers much better access to the small airways.

Disadvantages:
Compared to the rigid tube, the flexible tube may not have adequate diameter size to clear the airway of large foreign bodies, large mucus plugs, or obtain large biopsies.

Rigid Tube Bronchoscopy

Description:
A rigid open metal tube with a distal light, & a port for attaching oxygen or ventilation equipment. The tube is inserted into the mouth, through the trachea, & can go as far as the larger bronchi. To view segmental bronchi, a telescopic tube with mirrors is used.

Advantages over the flexible fiberoptic bronchoscope:
The large internal diameter of the tube allows aspiration of larger mucus plugs, & extremely thick inspissated secretions. Grasping forceps can be passed through the tube to remove larger foreign bodies & tumor biopsies.

Disadvantages:
The smaller airways cannot be accessed. The large rigid tube is extremely uncomfortable for a conscious patient. Normally this must be performed under general anesthesia, in the operating room. Performed by a surgeon or otorhinolaryngologist, with the assistance of an anesthesiologist.

Indications for Bronchoscopy:

Diagnostic
To evaluate problems with an endotracheal or tracheal tube.
To assess airway patency.
To examine airways for infiltrates, atelectasis, infection, disease, cancer, or abnormal tissue.
To obtain samples of abnormal tissue, secretions, cell washings, or biopsies for evaluation.
To examine airways for foreign material, or burn & smoke injuries.
To investigate possible causes of hemoptysis, wheeze, stridor, or sustained paroxysmal cough.
Therapeutic
To aid in difficult intubations.
To improve/ maintain airway patency by removing excessive viscous secretions, mucus plugs or a foreign body.
To cauterize an area, & destroy airway growths.
To deliver radiation & laser treatment.

Contraindications:

- Unstable medical condition, unstable hemodynamic or cardiac status.
- Uncorrected bleeding disorder.
- Refractory hypoxemia with inability to adequately oxygenate patient during procedure.
- Pregnancy (can be a contraindication, or physician may decide to proceed with caution).

Complications:

- Hypoxemia – Minimize by utilizing adequate oxygen.
- Bronchospasm – Most severe in asthmatics. Minimize by premedication with albuterol & Atrovent, & sedative.
- Hemodynamic changes (HR, BP, cardiac output) – Monitor closely & treat as necessary. Sometimes these changes are related to medication, or bronchoscopy technique & duration.
- If on mechanical ventilation – Increased PIP, decreased VT, inadvertent PEEP. Mainly related to the scope taking up appx. 50% of the airway tube radius. RT must monitor & adjust vent settings.
- Other complications w/ less than 3% occurrence rate – Vasovagal reaction, bleeding, cardiac dysrhythmia, pneumothorax, airway obstruction, nausea & vomiting, pneumothorax, pneumonia, fever, aphonia, respiratory arrest, & death (0.1%).

Important Pre-Procedure Preparation:

- Instruct patient about procedure, & obtain informed consent; unless emergency exists, & patient is unable to consent.
- Patient should not eat or drink for 4-12 hours prior to procedure.
- Dentures should be removed.
- Determine if patient has any allergies, bleeding disorders, or if taking anticoagulants.
- Assure patient is clinically stable & is not contraindicated as listed.
- Assure all equipment & supplies are prepared. See Supplies & Medications below. Assure machines are operating properly.
- Assure adequate personnel & facilities to safely perform procedure, & manage emergency situations. Procedure mandates the presence of bronchoscopy physician, & experienced assistants.

Supplies Needed:

- Bronchoscopy cart with all accessories & light source.
- Suction supplies, masks, gowns, gloves.
- Biopsy forceps, biopsy needles, brushes, bite blocks, sputum traps, specimen containers, microscopic slides, syringes, needles, saline, heparin, alcohol pads & cotton pads.
- Oxygen w/ multiple delivery devices, titrated to maintain adequate saturation. Resuscitation bag & mask. Pulse oximeter.

- Airway management & ventilation supplies & equipment. (Refer to chapter on Artificial Airways for a list of supplies.)
- Cardiac monitor & ECG leads.
- Fluoroscopy machine if requested by bronchoscopy physician.

Medications Utilized:

- Tranquilizer for anxiety – Benzodiazepines commonly used are Valium & Versed.
- Narcotic analgesic – morphine or fentanyl can be used to reduce pain & decrease laryngeal reflexes.
- Atropine to dry the airway & aid visibility. Atropine also reduces the risk of bradycardia, & hypotension.
- Bronchodilators – albuterol & Atrovent.
- Topical anesthetic – Cetacaine or lidocaine – applied to throat & nasal passages to numb the area, & reduce the gag reflex.
- Airway anesthesia using topical anesthetics, or nerve block. Lidocaine can be given by neb &/or direct instillation into airways.
- Syringes pre-filled with lavage solutions, saline, mucolytics, additional anesthetics, & vasoconstrictors.
- Epinephrine 1:1000.
- Additional Emergency Medication: resuscitative drugs, antiarrhythmics, narcotic antagonists like Narcan (naloxone), & IV fluids must be immediately available.

Bronchoscopy Physician Assistant's Duties:

- Aid in performing diagnostic & therapeutic procedures as listed.
- Deliver medications, & instill lavage solutions.
- Suction to keep airway clearly viewable.
- Continuously monitor patient's clinical status.

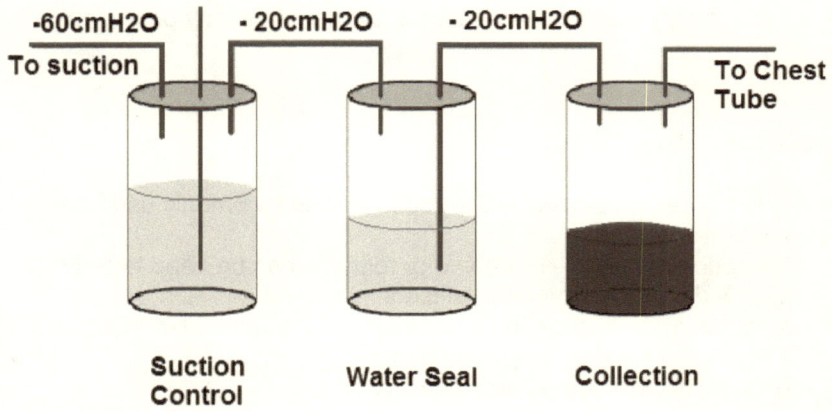

-60cmH2O   - 20cmH2O   - 20cmH2O

To suction   To Chest Tube

Suction Control   Water Seal   Collection

Chest tubes are long, clear, semi-stiff, plastic tubes that are inserted into the chest, so that they can drain collections of fluids or air from the space between the pleura. If the lung has been compressed because of this collection, the lung can then re-expand. All suction systems for chest tubes work in a very similar manner, though the appearance of the system varies.

The suction control bottle or chamber reduces the suction from the suction source to maintain steady suction on the chest tube. This is accomplished by the level of the water within, as opposed to the depth of the tube (in cmH_2O).

The water seal bottle or chamber isolates room air from the chest tube by creating a water seal. A Heimlich Valve can be used with small-bore chest tubes instead of the water seal bottle or chamber.

The collection bottle or chamber holds fluids removed from the patient via the chest tube. The collection bottle or chamber is graduated in order to measure the amount of fluid removed from the patient.

Indications: Pneumothorax (air), Hemothorax (blood), Fluid, Pleural effusion (usually serous, possibly from CHF, or a tumor process), Empyema (pus), & Sealant (antibiotic doxycycline).

Notes:
- Chest Tubes sizes range from a small bore 7 French to large bore 40 French.
- Chest tubes are equipped with Radiopaque stripes which, when placed into the pleural space to remove air or fluids, can be easily seen on the chest x-ray.
- Chest tubes are normally secured with sutures in order to prevent migration, & the insertion distance is recorded in order to check for tube migration. The chest x-ray is also used to check for migration.

How to Maintain & Troubleshoot Chest Tubes:

Maintain the water level in the suction control bottle/chamber at the specified level. This controls the suction for the system.

Maintain the water level in the water seal bottle/chamber above the bottom of the tube that connects to the collection bottle.

Assure the collection bottle/chamber is not allowed to fill completely. The only bottle/chamber routinely emptied is the collection unit. The water seal & suction control chambers or bottles should never be emptied when in use.

All levels should be indicated on the outside of each bottle or chamber.

Consistently check tubing for air leaks:
When using small bore tubes with a Heimlich Valve, leak testing can be accomplished by submerging the valve into water. Look for any back flow or bubbles after complete lung expansion, which would indicate a leak.

If the water seal bottle/chamber persistently bubbles you have a leak. You also have a leak if the skin sutured around the chest tube pulls away from the tube, & you hear a sucking sound. This can be corrected by occluding the hole with petroleum gauze. If the chest tube were to penetrate the lung the water seal bottle/chamber will bottle constantly.

Consistently check for bubbling:
If there is persistent bubbling or no bubbling at all in the water seal bottle/chamber, this can indicate malposition or migration. Chest tubes can become seated in the major fissure or become occluded, which will reduce their effectiveness. If there is no bubbling at all in the water seal bottle/chamber, this can also indicate clots or migration into tissue, which does not allow the chest tube to evacuate fluid. .

Removal:
Prior to removal, the most common practice to check for leaks is a 4-hour clamp test. If the condition requiring the chest tube returns, unclamp immediately & assess the cause of the leak. If the chest tube is removed 48 hours after leaks are no longer present, there is near 0% recurrence rate.

EKG Interpretation

- **EKG or ECG stands for electrocardiogram**. The "K" replaces the "C" to avoid confusion with the electroencephalogram or EEG. The EKG records the electrical activity of the heart.
- The heart contains electrically excitable special muscle cells & conductive tissue that are capable of generating an action potential without external stimulation. This is referred to as **automaticity**. As a result, the heart will **depolarize** & the cardiac muscle contracts (thus the pumping action of the heart). The heart will then **repolarize**. This electrical activity is displayed on the EKG as the **P, Q, R, S & T** waves.

EKG interpretation involves an evaluation of the following:

Rate, Rhythm, P wave, PR interval, QRS complex, QT interval, ST interval & axis. Also check for presence/ absence of hypertrophy, ischemia, & infarction.

Rate for Adults:

Normal	60 to 100 beats/min.
Bradycardia	Below 60 beats/min.
Tachycardia	Above 100 beats/min.

Rhythm:

Normal Sinus Rhythm (NSR)	A consistent, regular rhythm with a normal rate that originates in the SA Node (sinus node).
Arrhythmia	An irregular rhythm. Used to denote any variance from NSR.

P Waves: Check for presence or absence of a P wave before each QRS complex. Check for consistency of shape & size.

PR Interval: Normal is 0.12 to 0.20 seconds. Measured from the beginning of the P wave to the beginning of the QRS complex.

QRS Complex: Normal is 0.6 to 0.12 seconds. Measured from the beginning to the end of the QRS complex.

QT Interval: Normal is 0.32 to 0.44 seconds. Measured from the beginning of the QRS complex to the end of the T wave.

ST Interval: Depending on the lead, elevation can indicate infarction, depression can indicate ischemia.

Pacemakers of the Heart:

SA Node – Sino-atrial Node:
The SA Node is the normal pacemaker of the heart with an inherent rate of 60 – 100 beats per minute.

Ectopic Pacemakers:
If the SA Node fails as pacemaker of the heart, other Ectopic Foci may take over as they are potential pacemakers. This includes Atrial, Junctional, and Ventricular rhythms. Each of these Ectopic Foci have the ability to pace at a rate of 150-250 or higher in an emergency, & in certain pathological conditions.

Atrial Rhythm
The inherent rate of the atrial pacemaker is 60 – 80 beats per minute.

Junctional Rhythm
The inherent rate of the AV Junction or His Bundle is 40 – 60 beats per minute. This is also referred to as idiojunctional rhythm, & there is no P wave present.

Ventricular Rhythm
The inherent rate of the Ventricles is 20 – 40 beats per minute. This is also referred to as idioventricular rhythm.

Measurements & Rate Estimate:
Horizontally on an EKG strip, 1 small square equals 0.04 seconds. 1 large square equals 0.20 seconds (this contains 4 small squares). For a fast estimate of the rate, count the number of large squares between R waves.
Count downward at each box as in:

300, 150, 100, 75, 60, 50. This applies as long as the rhythm is regular.

EKG Time vs Amplitude

Time

0.04 Sec

0.20 Sec

1cm = 1mV   Amplitude

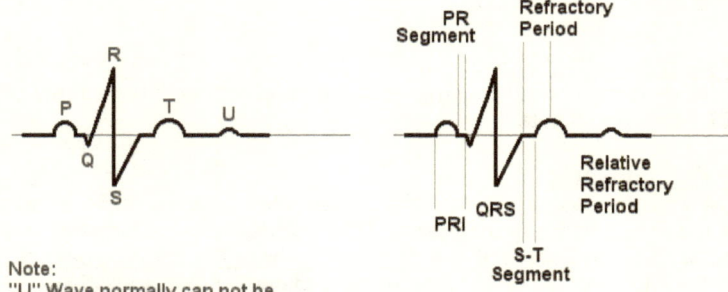

Note:
"U" Wave normally can not be seen on EKG due to next PRI.

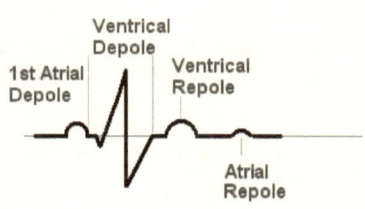

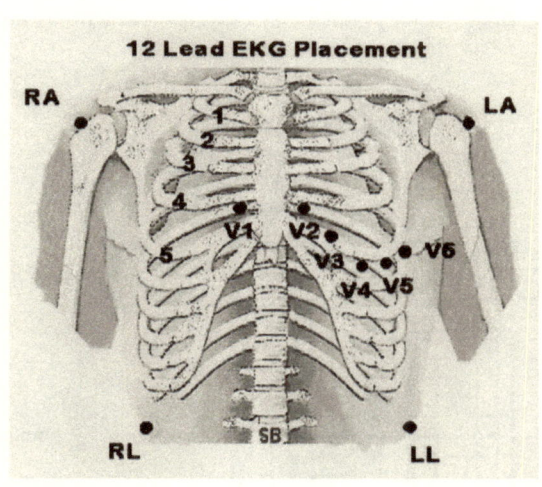

94

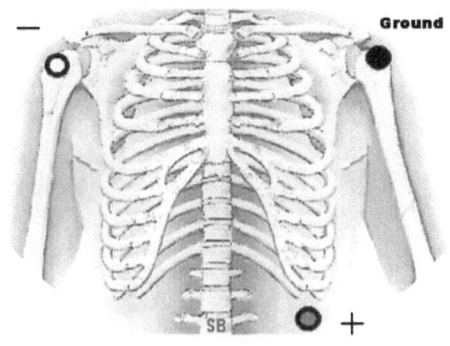

3 Lead EKG in Lead II

Normal Sinus

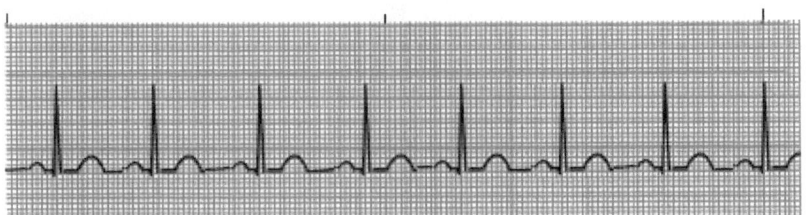

Sinus Bradycardia

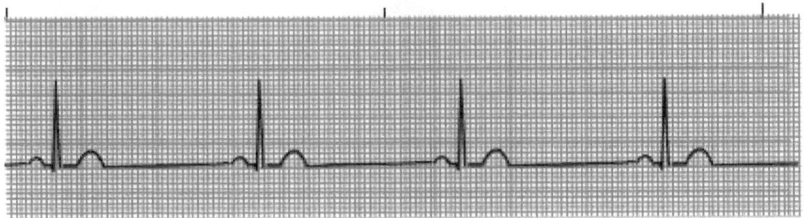

Sinus Tachycardia

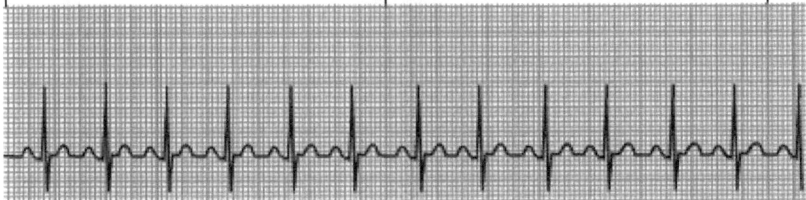

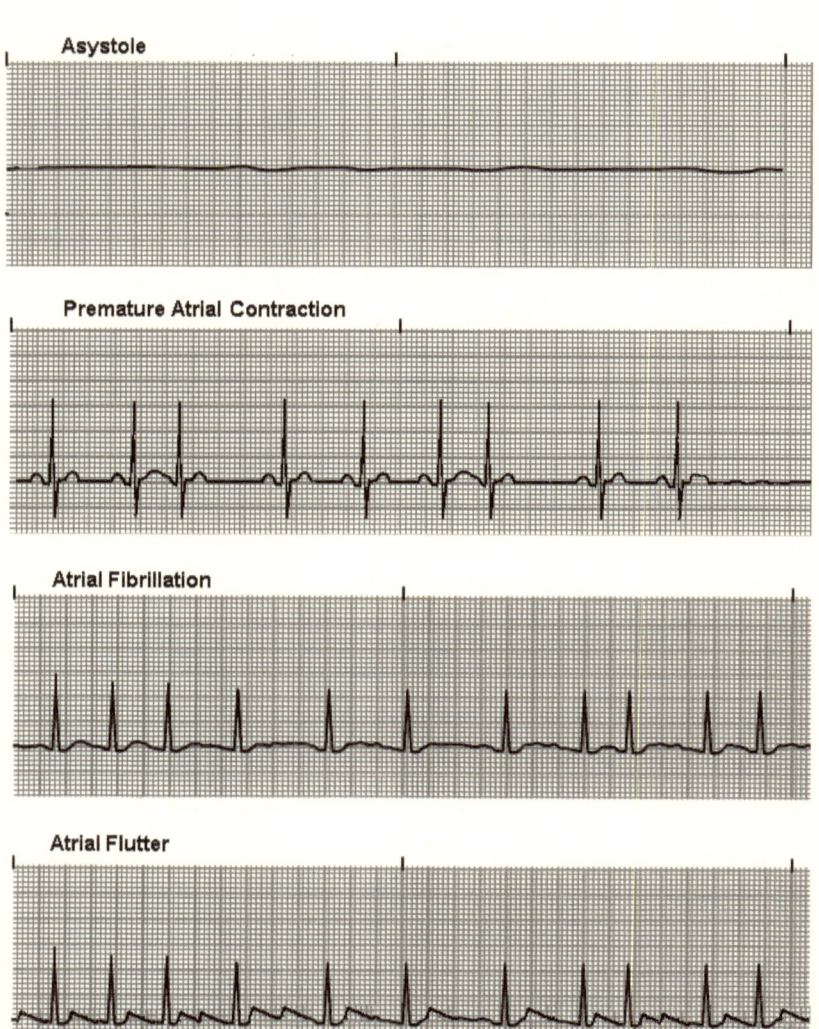

Asystole

Premature Atrial Contraction

Atrial Fibrillation

Atrial Flutter

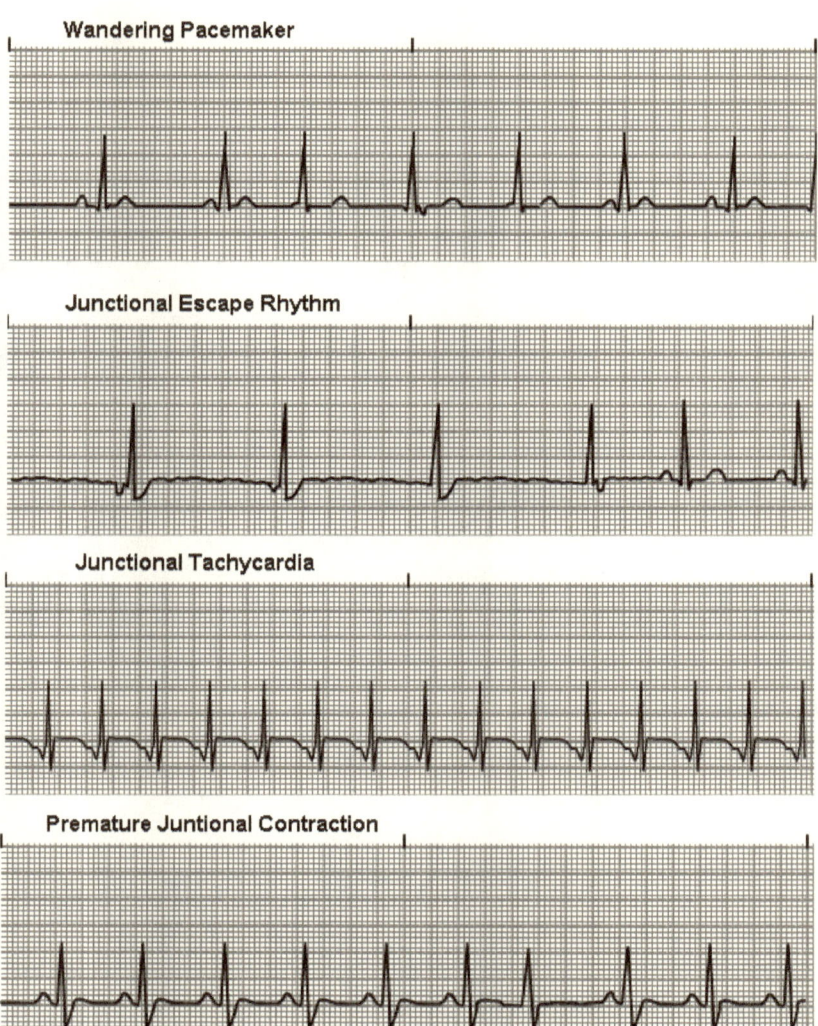

Wandering Pacemaker

Junctional Escape Rhythm

Junctional Tachycardia

Premature Juntional Contraction

97

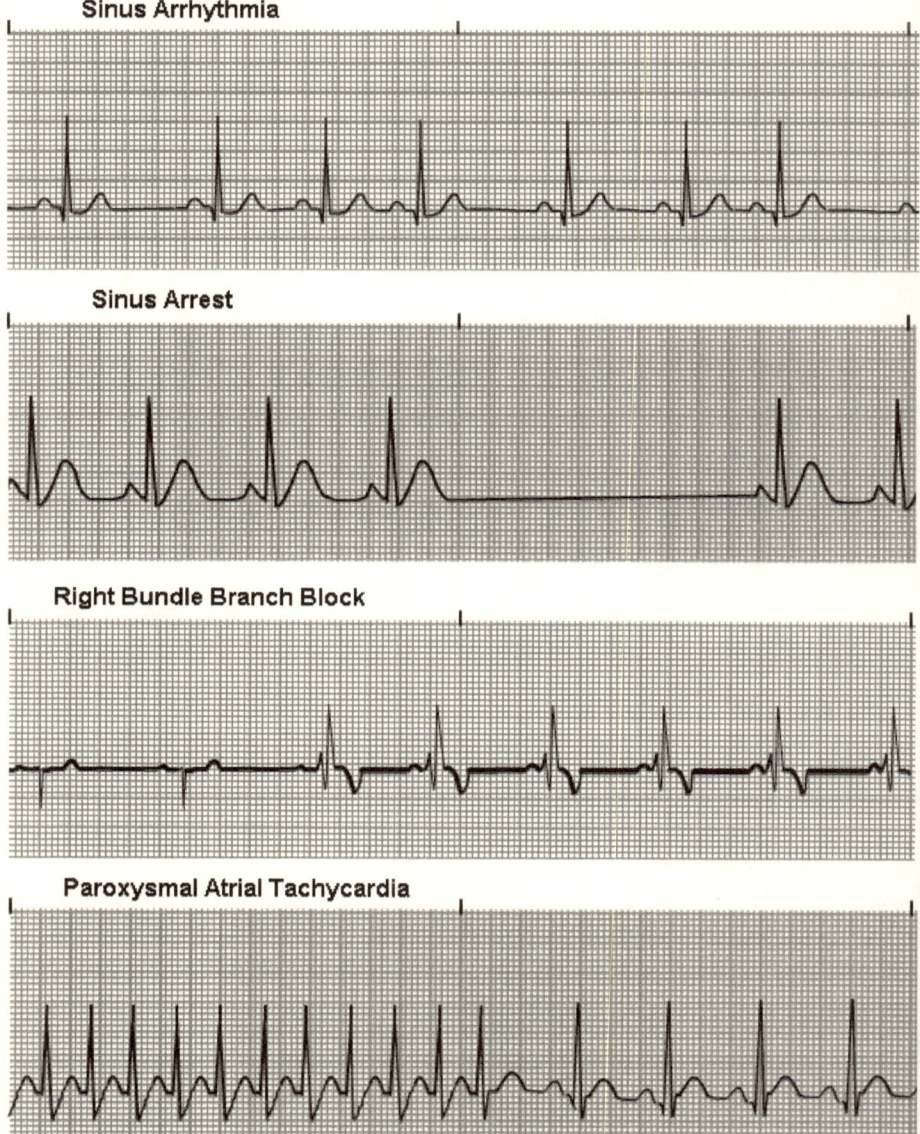

Sinus Arrhythmia

Sinus Arrest

Right Bundle Branch Block

Paroxysmal Atrial Tachycardia

First Degree A-V Block

Second Degree A-V Block (Motitz I - Wenckebach)

Second Degree A-V (Motitz II)

Third Degree A-V Block (Complete)

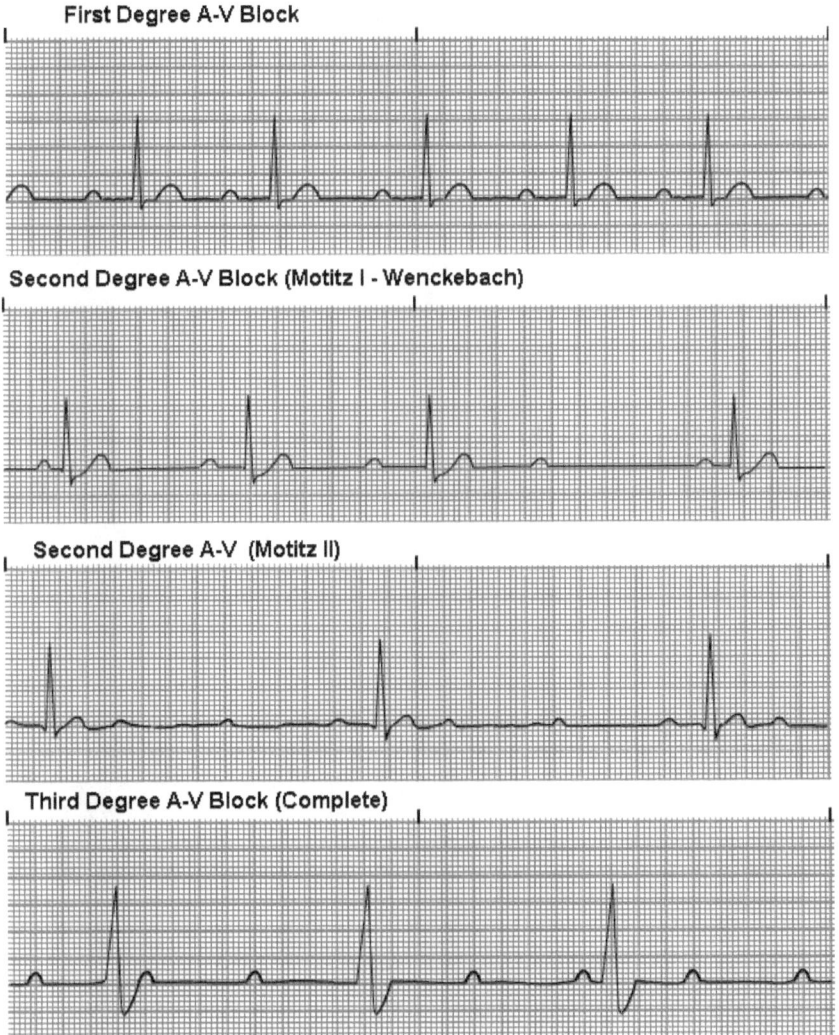

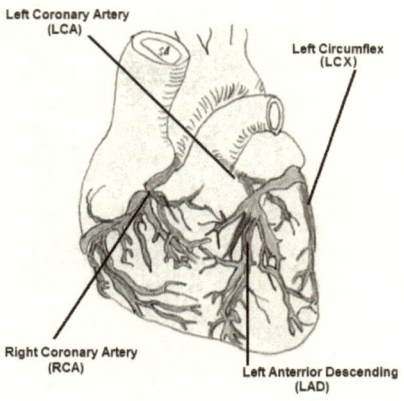

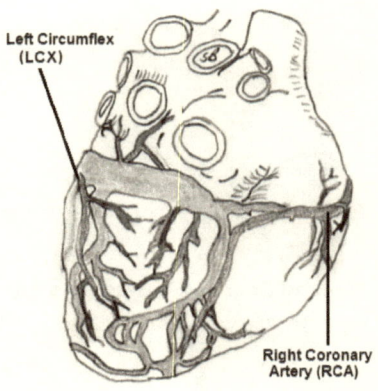

The LCA supplies the left atrium, left ventricle, septum, SA Node, Bundle of His, Right & Left Bundle Branches, anterior & posterior Hemibundles.

The RCA supplies the right atrium and right ventricles, SA & AV nodes, the Proximal Bundle of His, and the Posterior Hemibundle.

Normal 12 Lead EKG Traces

Normal 12 Lead EKG Traces

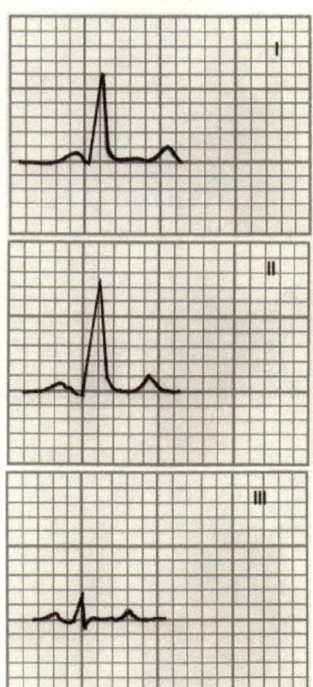

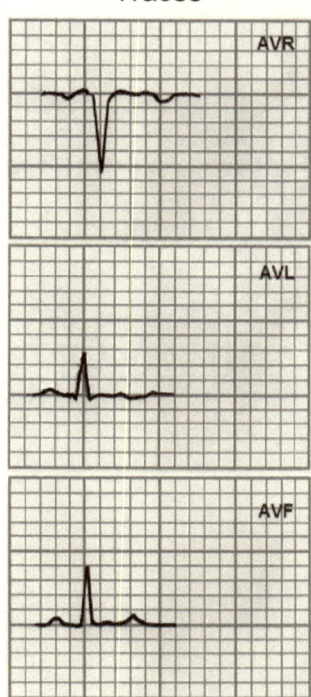

Normal 12 Lead EKG
Traces (Cont.)

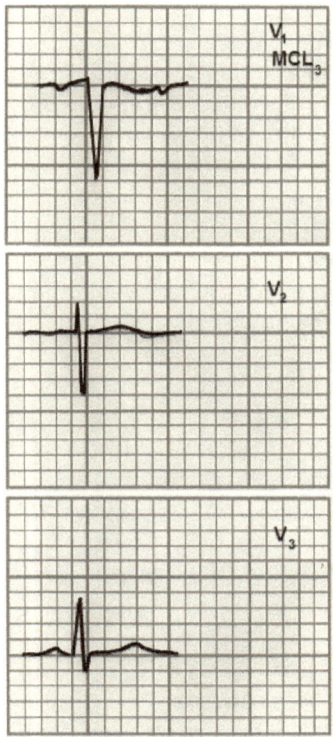

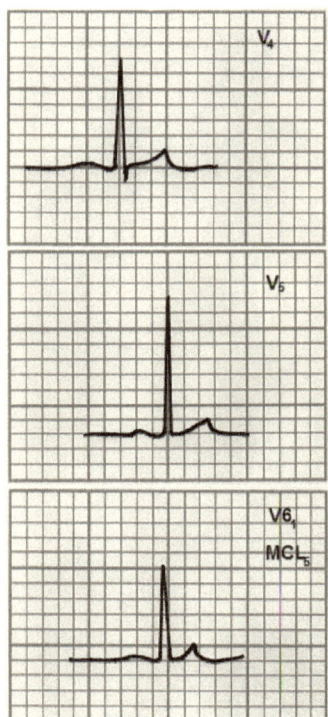

Myocardial Infarction (MI)

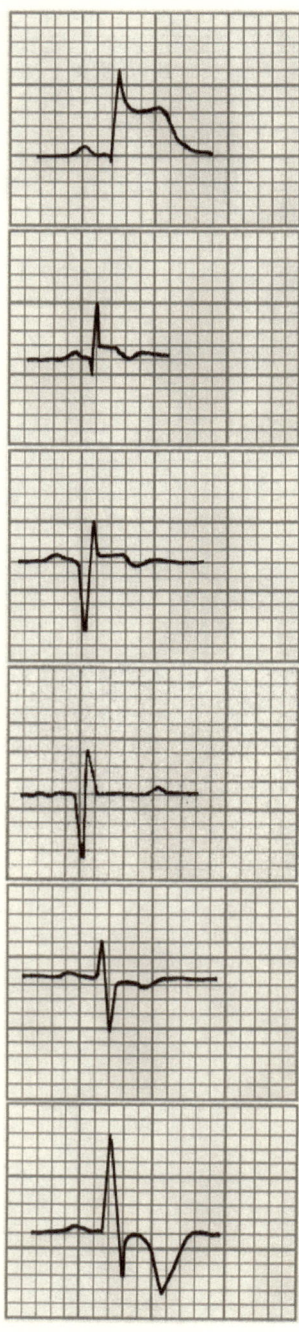

It is possible that changes in the EKG may not appear for hours after the event.

Injury
ST segments normally display elevation within minutes of the beginning of chest pain.

Ischemia
T waves fully invert within the first 24 hours.

Acute Infarction
Q waves $\geq$ 0.03 sec and/or 1/3 height or the QRS develop from 1 hour of the event.

Old Infarctions
Normal ST segments and the Q wave may be visible forever.

Reciprocal ST Depression
Will be found in leads away from the infarction.

Non Q Wave Infarction
ST segments will appear depressed and/or flat in 2 or more contiguous leads. There may be inverted T waves.

Bundle Branch Blocks

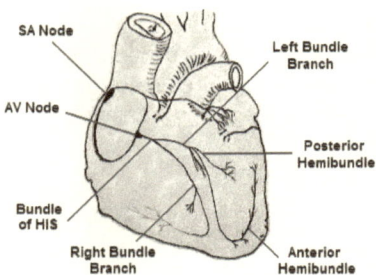

$QRS \geq 0.12sec$

If a Bundle Branch Block is evident on the rhythm strip it is more difficult to diagnose an MI.

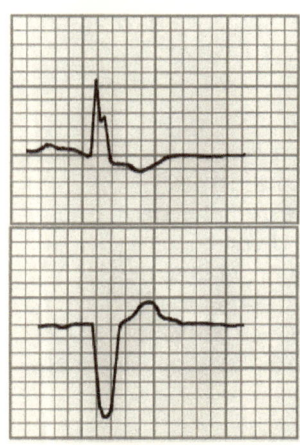

Left Bundle Branch Block
(LBBB) The R wave is notched, or has a rabbit ears appearance in I or V5 or V6.

Note the QS in lead V1, MCL1 Note: A rapid LBBB can appear to be ventricular tach in certain leads; be careful in diagnoses.

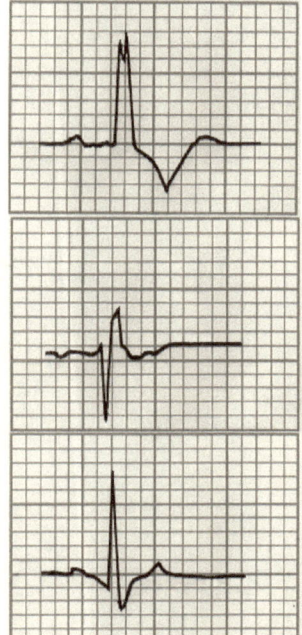

Right Bundle Branch Block
(RBBB) The R wave is notched, or has a rabbit ears appearance in V1 or V2. Also called R,S, R prime.

V1, V2, MCL1 can present like this.

Large S in 1, V5, V6

Electrolyte Drug Effects

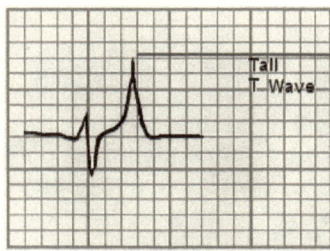

Hyperkalemia
Causes very tall peaked T waves, wide QRS and loss of P waves (in extreme cases).

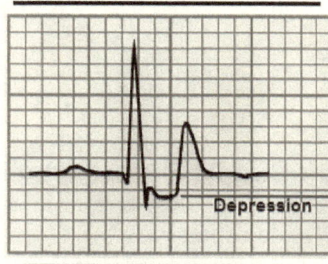

Digitalis Effect
Causes ST segments to be depressed & an asymmetrically inverted T wave.

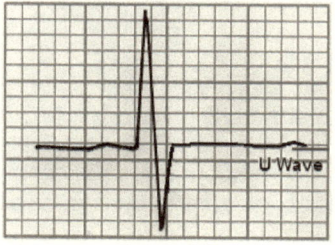

Hypokalemia
Causes a flat T wave and a large U wave.
Caution: Pt is at risk for VT, VF, or Torsades.
Caution: Hypokalemia exacerbates digitalis toxicity.

Chapter 13

PULMONARY FUNCTION TESTS

ALSO SEE: ABG's – Chapter 7
Mechanical Ventilation – Chapter 16
Misc. Formulas & Conversions – Chapter 20

Indications for PFT's:
→ Assess the pulmonary system
→ Test for the presence & extent of lung disease
→ Identify type of lung disease (obstructive vs restrictive)
→ Locate lung disease (small vs medium vs large airways)
→ Establish a course of therapy once dysfunction is identified.

PFT Chart Displays
Lung Volumes & Lung Capacities

TLC	FVC / VC	IC	IRV
			VT
		FRC	ERV
	RV		RV

Note: Lung Capacities contain two or more volumes.

Spirometry:

Spirometry **can** measure: FVC/VC, IC, IRV, VT, ERV, & flow rates.
Spirometry **cannot** measure: TLC, RV, or FRC. Why not? Because all
3 of these include the RV, which is air trapped in the lungs at the end of
a maximal exhalation. These 3 require special tests, which are detailed
later in this chapter.

NORMAL PFT VALUES (for Adults*)
+ CALCULATIONS & DEFINITIONS

* Normal values based on normal adults with ideal body weight (IBW).
Normal value calculation factors also include age, height, sex, & race.
Normal values decrease with age.

Volumes & Capacities	Normal Value Calculation	Normal Value for Female 62 Kg → 136 lb 168 cm → 66"	Normal Value for Male 75 Kg → 165 lb 178 cm → 70"
ERV	17 mL/kg	1054 mL	1275 mL
Expiratory Reserve Volume	Maximum volume of air that can be exhaled after a normal tidal exhalation.		
FRC	30 mL/kg	1860 mL	2250 mL
Functional Residual Capacity	Volume of air remaining in the lungs after a normal tidal exhalation. Includes RV air-trapping, which can't be measured by spirometry-requires special testing. ↑ FRC or ↑ RV/TLC ratio (> 20%) indicates an obstructive disorder. ↓ FRC & TLC indicates restrictive disorder.		
FVC or VC	65 mL/kg	4030 mL	4875 mL
Forced Vital Capacity or Vital Capacity	Maximum volume of air that can be expired after a maximal inspiration. FVC is equal in volume to VC, but the pt must exhale as quickly & forcefully as possible in order to assess flow rates.		
IC	50 mL/kg	3100 mL	3750 mL
Inspiratory Capacity	Maximum amount of air that can be inspired after a normal tidal exhalation.		
IRV	40 mL/kg	2480 mL	3000 mL
Inspiratory Reserve Volume	Maximum volume of air than can be inhaled after a normal tidal inspiration.		

RV RV/TLC ratio	16 mL/kg or 20% of TLC	992 mL	1200 mL
Residual Volume	Volume of air remaining in the lungs after a maximal exhalation. Includes air-trapping, which can't be measured by spirometry- requires special testing. ↑ FRC or ↑ RV/TLC ratio (> 20%) indicates an obstructive disorder. ↓ FRC & TLC indicates restrictive disorder.		
TLC	80 mL/kg	4960 mL	6000 mL
Total Lung Capacity	The total volume of air contained in the lungs.		
VT or TV	7 mL/kg	435 mL	525 mL
Tidal Volume	Volume of air that is inhaled & exhaled during normal quiet breathing.		

Flow Rates & Time%	Normal Value Calculation	Normal Value for Female 62 Kg → 136 lb 168 cm → 66"	Normal Value for Male 75 Kg → 165 lb 178 cm → 70"
FEV 0.5 seconds FEV 0.5/%FVC	60% of FVC	2420 mL	2925 mL
Forced Expiratory Volume in first 0.5 seconds of FVC. 0.5 Time% normally 60% of FVC.			
FEV 1.0 seconds FEV 1/%FVC	80% of FVC	3225 mL	3900 mL
Forced Expiratory Volume in first 1 second of FVC. 1.0 Time% normally 80% of FVC.			
FEV 2.0 seconds FEV 2/%FVC	94% of FVC	3790 mL	4580 mL
Forced Expiratory Volume in first 2 seconds of FVC. 2.0 Time% normally 94% of FVC.			
FEV 3.0 seconds FEV 3/%FVC	97% of FVC	3910 mL	4730 mL
Forced Expiratory Volume in first 3 seconds of FVC. 3.0 Time% normally 97% of FVC.			
FEF 200-1200 or MEFR	6 L/sec	360 L/min	360 L/min
During the FVC: Forced Expiratory Flow exhaled between 200 mL & 1200 mL. Also called Maximum Expiratory Flow Rate. Measures function of the large airways.			

FEF 25-75% or MMFR	4.7 L/sec	280 L/min	280 L/min

Forced Expiratory Flow rate over the middle 50% of FVC. Also called Maximal Mid-Expiratory Flow Rate. Measure function of the small & medium airways.

Other Pulmonary Mechanics	Normal Value Calculation	Normal Value for Female 62 Kg → 136 lb 168 cm → 66"	Normal Value for Male 75 Kg → 165 lb 178 cm → 70"
IS	50 mL/kg	3100 mL	3750 mL

Incentive spirometry is used to measure inspiratory respiratory muscle strength, improve lung aeration, & prevent atelectasis.

MEP	>= + 80 cmH$_2$O	>= + 80 cmH$_2$O	>= + 80 cmH$_2$O

Maximal Expiratory Pressure. Assess respiratory muscle strength. Pt inhales deeply & expires all air into device, while peak expiratory pressure is measured.

MVV or MBC	150-170 L/min	150 L/min	170 L/min

Maximum Voluntary Ventilation or Maximum Breathing Capacity. Measures respiratory muscle strength, compliance, & resistance. Pt breathes as fast & deep as possible for 12-15 seconds into spirometer w/accumulator recording. MVV exaggerates air-trapping. Very sensitive & can give indication of an obstructive disease in the early stages. Very effort dependent.

NIF or MIP	−80 to −100 cmH$_2$O	−80 to −100 cmH$_2$O	−80 to −100 cmH$_2$O

Negative Inspiratory Force is used to assess respiratory muscle strength. Pt inhales maximally while peak inspiratory pressure is measured. Same as MIP (Maximal Inspiratory Pressure).

PF Peak Flow or PEFR	400 – 600 L/min	400 L/min	600 L/min

Peak Flow or Peak Expiratory Flow Rate. Pt inhales deeply & blows all air into device as fast as possible.

PFT SPECIAL TESTS

Angiogram/Arteriogram (See Pulmonary Angiogram/Arteriogram)

Body Plethysmograph / Body Box / Thoracic Gas Volume (TGV):
Most accurate method for measuring FRC, RV, & TLC. Measures
total TGV, including air trapped in smallest airways. Pt pants at FRC
against closed shutter, at about 2 breaths per second, while
pressures & volumes are measured.
Normal Value: Pt to achieve predicted values for TGV, TLC & FRC.
Abnormal Results: TGV ↑ in obstructive disease, ↓ in restrictive
disorder.

DLCO Lung Diffusion for Carbon Monoxide or Gas Diffusion Testing:
Measures factors that affect diffusion of air across A-C membrane,
detecting if surface area for diffusion is reduced. Pt inhales small
concentration of carbon monoxide, helium & air. Pt exhales into
device that analyzes gas concentrations & calculates DLCO.
Normal Value: Single breath DLCO 25 mL/minute/mmHg.
Steady state DLCO 17 mL/minute/mmHg.
Abnormal Results: DLCO ↓ in emphysema, pulmonary fibrosis,
embolism, & sarcoidosis.

Flow-Volume Loop:
Measures the volumes & flow rates of the vital capacity. Exp flow is
above baseline & insp flow is below baseline. Pt inhales as deep as
possible, then exhales forcefully for the FVC, then inhales maximally
again. A graphic loop as displayed.
Normal Value: Pt to achieve predicted volumes & flow rates.
Abnormal Results: Can detect obstructive & restrictive patterns, ↓
volumes, ↓ flow rates, airway resistance & small airway disease.

Helium Dilution (closed circuit):
Method of measuring the FRC, then calculating RV & TLC. Not quite
as accurate as plethysmograph for detecting trapped air. Pt
breathes mixture of air with 10% helium, until equilibrium takes
place at about 5 -7 minutes.
Normal Value: Equilibrium at <= 7 minutes. Predicted FRC.

Abnormal Results: If equilibrium takes longer (up to 20 min), it indicates obstructive disease. FRC is ↑ in obstructive disease, ↓ in restrictive disorder.

Nitrogen Washout (open circuit):

Method of calculating the FRC & RV. Not quite as accurate as plethysmograph for detecting trapped air. Also measures evenness of distribution of ventilation, w/breath-by-breath curve. Pt breathes 100% O_2 for 2-7 minutes, exhaling all gas into an analyzer, until nitrogen remaining in lungs is less than 2.5%. Then pt exhales completely. Fractional concentration of alveolar nitrogen (FAN_2) is measured, & FRC is computed.

Normal Value: Less than 8 minutes to reach less than 2.5% nitrogen in lungs. Pt to achieve predicted FRC.

Abnormal Results: Greater than 7 minutes to reach 2.5% nitrogen remaining in the lungs indicates poor distribution of ventilation, obstructive disorder, or possible pulmonary embolism. FRC is ↑ in obstructive disease, and ↓ in restrictive disorder.

Pulmonary Angiogram or Arteriogram:

Measures blood flow distribution. A radiographic study of the arteries is performed after injecting radiopaque dye. Still pictures & motion pictures are taken, with observation of blood flow through blood vessels. Useful in detecting unperfused blood vessels & pulmonary embolism.

SBN_2 Single Breath Nitrogen Elimination w/ Closing Volume (CV) & Closing Capacity(CC):

Measures the evenness of distribution of inhaled gases. Pt exhales maximally, inhales 100% O_2 maximally, then slowly exhales the gas until lungs feel empty. The exhaled gas passes through an N_2 analyzer that measures the change in concentration of nitrogen. The first 750 mL of air exhaled is mostly deadspace, & is discarded (phase I & II). The next 500 mL of exhaled air (phase III) is used to measure distribution.

Normal & Abnormal Values:

Phase III N_2 rise should be less than 1.5%. A phase III N_2 rise > 1.5% indicates uneven distribution of ventilation or uneven flow rates, with possible pulmonary embolism.

CV is Phase IV, & is very sensitive for detecting early airway closure. CV should = 10-20% of VC. The CV% is ↑ in small airway obstruction.

CC is Phase V. CC should = 30-40% of TLC.

Volume of Isoflow (VisoV) & Vmax50:

Test consists of 2 maximal expiratory flow curves. The 1st FVC uses air. The 2nd FVC uses O_2 20% + helium 80%. Volume remaining in lungs after 2nd FVC is the volume of isoflow. Vmax50 measures flow at 50% of the VC, similar to FEF 25-75%..

Normal & Abnormal Values: The Volume of Isoflow is normally 10-20% of VC. VisoV is ↑ in small airway disease. The Vmax50 can detect changes in airway resistance in small & medium airways.

V/Q scan or Ventilation/perfusion scan:

Measures gas & blood flow distribution (ventilation & perfusion). Radiolabeled gas (xenon) is inhaled. Then radioisotope is injected. This is followed by a study of mismatches between ventilation & perfusion. The V/Q scan can detect poorly ventilated areas, unperfused blood vessels, & pulmonary embolism.

PFT INTERPRETATION

PFT Interpretation:
Based on Percent of Predicted Value

Normal.........................80-120% of predicted value.
Mild disorder.................65-79% of predicted value.
Moderate disorder........50-64% of predicted value.
Severe disorder............Less than 50% of predicted value.

Post Bronchodilator % Improvement:

Non-significant improvement.......Less than 15%
Significant improvement..............15% or greater
Very large improvement may be indicative of asthma.

Disease Patterns on PFT's: Restrictive vs Obstructive

PFT Value	Restrictive Disorder	Obstructive Disease
ERV	N or ↓	N or ↓
FRC	N or ↓	↑
FVC or VC	↓	N or ↓
IC	N or ↓	N or ↓
MVV or MBC	N or ↓	↓
IRV	↓	N or ↓
PF Peak Flow	N or ↓	↓
RV	N or ↓	↑
TLC	↓	↑
VT	N or ↓	Varies
FEV 0.5 seconds	N or ↓	↓
FEV 1.0 seconds	N or ↓	↓
FEV 2.0 seconds	N or ↓	↓
FEV 3.0 seconds	N or ↓	↓
FEF 200-1200	N or ↓	↓
FEF 25-75%	N or ↓	↓

N = Normal

Obstructive disease pattern:
Decreased flow rates, increased RV, increased FRC, increased TLC.

Restrictive disease pattern:
Decreased volumes, decreased TLC.

Obstructive pulmonary diseases include asthma, bronchitis, bronchiectasis, emphysema, cystic fibrosis, & bronchopulmonary dysplasia (BPD). Most other pulmonary dysfunctions are restrictive.

PFT Notes

ATS Standards: Protocols that help prevent procedural errors, thereby assuring accurate results.
→ PFT Equipment must be cleaned, calibrated, & maintained on a regular schedule.
→ Patients must be instructed prior to the test, & given 3 attempts when appropriate (e.g. FVC).
→ Inform the pt that many PFT's are very effort dependent. Help motivate the pt to put forth the best possible effort.

ATPS: Spirometry measures volumes at ambient temperature, pressure, & saturated (ATPS) conditions.
BTPS: The measurements are then adjusted for the temperature difference between the spirometer & body temperature, pressure, & saturated (BTPS) conditions.

Specialized Imaging Procedures

CT Scan / CAT Scan

Computed tomography (CT), or computed axial tomography (CAT), is an advanced x-ray procedure using tomography. Digital geometry is used to generate cross-sectional views & 3-dimensional images of the internal organs & structures of the body, in order to locate abnormalities. CT is also used in procedures to accurately guide the placement of instruments or treatments. A contrast material is sometimes used. CT is excellent for detecting acute & chronic changes in the lungs. For example: pulmonary embolism (PE), pneumonia, cancer, emphysema, fibrosis, & lymphadenopathy.

MRI (Magnetic resonance imaging)

MRI is a test that uses a magnetic field and pulses of radio wave energy to obtain images of organs and structures inside the body. A contrast material may be used to obtain more clearly defined images. Often, MRI gives information that cannot be seen on X-ray, ultrasound, or CT scan. MRI is useful in finding problems within organs. Also locating or diagnosing: tumors, cancer, bleeding, aneurysm, injury, stroke, blood vessel problems, nerve problems, spinal problems, blockage, infection, & much more.

ALSO SEE: PFT's for specialized radiograms –
***   Angiogram, Arteriogram, V/Q Scan***

X-ray Projections

Apical Lordodic view: A projection of the lung apices.

(A-P) Anterior-posterior: X-rays pass through the patient from front to back, with the patient's back placed on the film. This is the typical portable chest x-ray used for the bedridden, critically ill, or difficult-to-move patient.

(P-A) Posterior-anterior: X-rays pass through the patient from back to front, with the patient's chest placed on the film. P-A is preferable over the A-P projection as it offers a better view, but it's more difficult to obtain with bedridden or critically ill patients.

Lateral view & Lateral Decubitus: A side to side view. The lateral decubitus can be used to diagnose a pleural effusion, as the fluid can be visualized as "layering out". The lateral neck x-ray is used for diagnosing epiglotittis, croup, or foreign bodies in the neck area.

Oblique position: A slanting diagonal view that aids in localizing lesions.

High Density & Low Density Structures

The x-ray consists of high-density & low-density structures.

- High-density structures appear white (radiopaque or radiodense) on an X-ray because more of the x-ray is absorbed, & less of the x-ray reaches the film. The higher the density, the whiter it appears. The bones are the most dense naturally-occurring density, followed by fat, then water.

- Low-density structures appear dark (radiolucent) on x-ray because less of the x-ray is absorbed, & more of the x-ray reaches the film. Normal lungs appear radiolucent because they contain air, which has the least density of all.

Normal Anatomic Structures

Bones: The clavicles, ribs, & sternum are easily located. These make good landmarks for locating other structures. The inferior angle of the scapula is located at the 8th rib posteriorly. If the patient is inhaling maximally, the 10th or 11th rib should be visible. The ribs can be examined for any fractures. The contour of the spine can be studied for any abnormal curvature.

Diaphragm: The right hemidiaphragm is dome shaped & higher than the left due to the presence of the liver under the right, & the heart pushing down on the left. The anterior portion of the liver is near the 5th rib. The upper domes of the diaphragm should be smooth & rounded.

Heart: The heart is predominantly on the left side of the sternum. The right heart border creates a slight bulge on the right side of the spine. The cardiothoracic ratio (C/T ratio) should be less than 50%. The ratio describes the total size of the heart in relation to the total size of the thorax. The aortic arch & ascending aorta are located at the level of the 2nd rib at the junction of the manubrium & the body of the sternum. The left pulmonary artery can be seen below the aortic arch.

Hilum: The hilar regions on both sides of the sternum include the pulmonary arteries, veins, & some segmental bronchi. The left hilum is normally slightly higher than the right. Some small round radiopaque densities are normal in the hilar area as these represent blood vessels seen on end.

Lung Parenchyma: Vascular markings are normal in the lungs & should be seen throughout both lung fields. Vascular markings are the lung tissue, blood vessels & lymphatics. The carina can be visualized at the point of bifurcation of the right & left mainstem bronchus at the level where the 2nd rib joins the sternum. The trachea is easily located, and on an intubated or trach patient, the tip of the tube should be positioned at 2 - 3 cm above the carina.

Mediastinum: The area between the lungs. Includes the heart, blood vessels, trachea, main bronchi, lymphatics, nerves & connective tissue. A mediastinal shift indicates an abnormality as listed in next section.

Pleural Surface: Surrounding the perimeter of the lungs are the visceral pleura & parietal pleura with a potential space in between. The visceral pleura covers both lungs & lines the interlobar fissures. The parietal pleura lines the inside of the chest wall & covers the sides of the pericardium & the hemidiaphragms. The two pleura fuse together in the area of the hila. The pleural space normally cannot be viewed on the x-ray, but if fluid or air enters the pleural space it can be seen. Both costophrenic angles should come to a sharp point, indicating there is no fluid in the pleural space.

Soft Tissues: Tissues, muscle, and fat in the chest and neck area. One can check for subcutaneous emphysema in the soft tissues.

Abnormalities in the Chest X-ray

X-ray Terminology	View- Viewable Characteristics on the x-ray - Dx- Possible Pathologies/Diagnoses
Air Bronchograms	View- Bronchi are dilated, bulbous, outlined. Dx- Bronchiectasis, IRDS stage 3, pulmonary edema, pulmonary hemorrhage, or pneumonia. Bronchography with the injection of a contrast medium aids in diagnosis.

A-P Diameter Increased	See Hyperinflation.
Blunting	See Costophrenic Blunting.
Butterfly pattern	View- Whitish butterfly pattern. Dx- Pulmonary edema.
Consolidation	View-Solid white appearance. Dx- Typically indicates pneumonia, but can also indicate pleural effusion, tumor, or other solid abnormality.
Costophrenic Blunting	View- Obliterated/rounded angle or meniscus w/basilar infiltrates. Dx- Pleural effusion. (Lateral decubitus x-ray aids in diagnosis.)
C/T Ratio increased, or C/T greater than 50%	Ratio describes the total size of the heart in relation to total size of the thorax. Dx- C/T Ratio greater than 50% indicates cardiomegaly, heart disease, CHF, or COPD.
Diaphragm with Flattened Domes	Dx- Indicative of hyperinflation & COPD.
Diffuse	Spread bilaterally throughout the lungs.
Fluffy infiltrates, diffuse	View- Whitish fluffy diffuse pattern. Dx- Pulmonary edema.
Ground Glass pattern	View- Whitish ground glass pattern. Dx- ARDS or IRDS.
Heart abnormalities Also see C/T Ratio	X-ray can aid in diagnosing right heart failure (cor pulmonale) or left heart failure (CHF).
Hemidiaphragms – one significantly higher than other	Dx- Possible unilateral phrenic nerve paralysis.
Hilar Region Enlarged	Dx- Can be indicative of lymphoma, sarcoidosis, or engorged pulmonary vessels.
Honeycomb pattern	View- Whitish diffuse honeycomb appearance. Dx- ARDS, IRDS, interstitial fibrosis or edema.
Hyperinflation or Increased A-P Diameter	View- Dark appearance representing air. Dx- Excessive air indicative of COPD, emphysema.
Hyperlucent or Increased Radiolucency	View- Dark appearance representing air. Dx- Increased air indicative of: bullae, pneumothorax, COPD, or emphysema.
Infiltrates	View- White appearance. An infiltrate is any ill-defined radiodensity or opacity. Dx- Can indicate

	multiple pathologies as listed here: atelectasis, pneumonia, edema, or other pathology. Infiltrates can be patchy, platelike, solid, localized, fluffy, or diffuse. See separate listings.
Kerley B Lines Perpendicular lines	Dx- Interstitial edema or edema due to left heart failure.
Mediastinal Shift	Dx- Shift away from affected side in pneumothorax. Shift toward affected side in atelectasis.
Meniscus	See Costophrenic Blunting.
Miliary pattern	Dx- Alveoli filled with fluid.
Obliterated/rounded angle or meniscus	See Costophrenic Blunting.
Opacities	White appearance. Normal for bones & organs. Otherwise, Dx- Infiltrates, atelectasis, pneumonia, edema, pleural effusion, or other pathology.
Patchy or Platelike Infiltrates	Dx- Probable atelectasis.
Pulmonary Arteries Dilated	Dx- Pulmonary hypertension.
Radiopacity Radiopaque Radiodensity	View- White appearance. Normal for bones & organs. Otherwise, Dx- Infiltrates, atelectasis, pneumonia, edema, pleural effusion, or other pathology.
Radiolucent	View- Dark appearance representing air. Normal for lungs. (Also see Hyperlucent.)
Reticular pattern	View- White streaks. Dx- Interstitial infiltrates or fibrosis.
Reticulogranular pattern	View- Diffuse whitish ground glass appearance. Air bronchograms may also be present. Dx- ARDS or IRDS.
Soft Tissue w/ Radiolucent Streaks	Dx- Subcutaneous emphysema.
Spinal Curvature	Best viewed on lateral projection. Dx- Abnormal curvature indicative of kyphosis, scoliosis, kyphoscoliosis, humpback, bone deformities related to aging, or other spinal defects.
Vascular Markings	Normal for lymphatics, lung tissues, & blood vessels. Otherwise, Dx- Increased vascular markings = Pulmonary edema, fibrosis, or other pathology. Absent peripheral vascular markings = Pneumothorax.

Chapter 15

**ARTIFICIAL AIRWAYS
& SUCTIONING**

Intubation & Extubation Guidelines
Oral & Nasal Airways
ET Tubes w/ Tube & Blade Sizes, Cuff Pressures
Notes: Air Cuff, Foam Cuff, Cuffless
         PVC Tubes, Fenestrated Tubes & Metal Tubes
EOA, PMV
Suctioning Guidelines & Vacuum pressures

INTUBATION Guidelines:

- Prepare all equipment including:
  Several tubes, syringe, stylet, laryngoscope blades, tape or other tube-securing device, Magill forceps, oral airways, bite block, lubricant, lidocaine, racemic epinephrine, suction equipment, $ETCO_2$ detector, resuscitation bag & mask. Check the cuff. Set up a ventilator on standby. Nursing should prepare common intubation medications (neuromuscular blockers, sedatives, etc).
- Premedicate patient per protocols.
- Lubricate tube, & use lidocaine or Cetacaine to provide local anesthesia for oral or nasal cavities. Racemic epi can be used to provide vasoconstriction for nasal intubation.
- Hyperoxygenate & ventilate the patient prior to intubation, & maintain adequate oxygenation & ventilation throughout procedure.
- Limit intubation attempt to a maximum of 15 seconds.
- Watch for hazards & complications as listed below in "Suctioning".
- For oral intubation- Position patient's head properly & choose appropriate size curved or straight blade. Use stylet to shape & guide the ET tube. Visualize the ET tube passing through cords.
- For nasal intubation- One can use Magill forceps to guide ET tube, or one can do a blind intubation.

- Once intubated, inflate the cuff & secure the tube.
- Assess for successful intubation:
  Visible moisture condensate appearing in tube.
  Equal bilateral breath sounds & lung expansion.
  No increase in abdominal distention.
  Assess exhaled CO_2 with $ETCO_2$ detector (optional).
  Verify proper placement of tube at 2-3 cm above the carina with a chest x-ray.

EXTUBATION Guidelines:

Assure patient meets extubation criteria. (See Mechanical Ventilation chapter.)
- Explain the procedure to the patient.
- Preoxygenate & ventilate the patient.
- Suction the trachea & oropharynx.
- Remove tape or other tube securing devices.
- Deflate the cuff.
- Remove the cuff at end inspiration while patient coughs.
- Suction the oropharynx.
- Place patient on oxygen.
- Assess for dyspnea, hypoxemia, or stridor, & treat as indicated.
- Monitor patient's vital signs & clinical status closely.
- Reassess & monitor patient for adequate oxygenation & spontaneous ventilation.

ORAL & NASAL AIRWAYS:

ORAL PHARYNGEAL AIRWAY:
Used to maintain a patent oral airway in an unconscious patient. Not well tolerated by a conscious patient. Also used as a bite block & to facilitate oral suctioning. Proper size for patient is from the tip of the chin to the angle of the jaw. Insert upside down until tip is at the end of the tongue, then rotate 180° into proper position.

NASAL PHARYNGEAL AIRWAY / NASAL TRUMPET:
Used to maintain a patent nasal airway in a conscious or unconscious pt & to facilitate deep tracheal suctioning. Diameter should be slightly less than pt's nare. Lubricate for easier placement. Must be periodically changed with a clean, lubricated nasal trumpet.

ET Tube & Blade Sizes
& Suction Catheter Sizes

Age	ET Tube Size Internal Diameter (mm)	Laryngo-scope Blade Size	Suction Catheter (French size)
Newborn < 1000 grams	2.5	0	5
1000-2000 grams	3.0	0	6

2000-3000 grams	3.5	1	6-8
> 3000 grams	3.5 – 4.0	1	6-8
6 months	3.5 – 4.0	1-2	6-8
1 year	4.0	1-2	8
2-3 years	4.5 – 5.0	2	8
4-7 years	5.0 – 6.0	2	8-10
8-10 years	6.0 – 6.5	2-3	10
12 years	7.0	2-3	10
15 years	7.0 – 7.5	3	10-12
Adult Female	7.5 – 8.5	3-4	12
Adult Male	8.0 – 9.5	3-4	14

Calculating proper suction catheter size:
  (Internal Diameter mm of ET tube / 2) x 3 = French size catheter

Cuff Pressures

Tube Cuff pressure should always remain below 20 mmHg (26 cmH$_2$0) to prevent edema & necrosis.

Notes: Air Cuff, Foam Cuff, Cuffless
PVC Tubes, Fenestrated Tubes & Metal Tubes

• Most tubes have a standard 15 mm adapter on the proximal end. (An exception is the metal tubes that require an adapter.)

   ET Tubes are made of PVC (polyvinylchloride), and have an air cuff or a foam cuff. However, the smallest tubes (< 5 mm) are uncuffed due to the very small airways of infants & children. With the air cuff, the pilot tube must be closed to seal it, & one can view the pilot balloon to see if the cuff is inflated w/ air. On foam cuffs the pilot tube is left open to air.

   Tracheal tubes come with or without a cuff depending on application. For a spontaneously breathing patient, the choice in tracheal tubes is the regular PVC tube – cuffed or cuffless, a fenestrated tube, or a metal tube. For positive pressure ventilation for adults, a cuff is needed.
   The fenestrated tracheostomy tube is a double-cannula tube, with the outer cannula having an elongated fenestration (hole) in the outer curve. This can be used to facilitate weaning from a trach tube as the pt

is forced to breathe through the upper airway when the inner cannula is removed, and the cuff is deflated, and the outer cannula is plugged.

Metal tubes are cuffless, & are usually used for long-term tracheostomy patients who do not need positive pressure ventilation.

EOA (Esophageal Obturator Airway)

Description:
The EOA is a tube that is inserted into the esophagus, not into the trachea. The distal end is plugged, & there are 16 holes at the level of the oropharynx. A mask clips onto the proximal end of the tube, thus allowing ventilations to be delivered to the lungs by a resuscitation bag. The EOA cuff is inflated to block air from entering the stomach, & to prevent aspiration of gastric contents. It provides better ventilation than bag-&-mask since it blocks air from entering the stomach, however, it is not better than intubation.

Utilization:
The EOA is designed for short-term use only (up to 2 hours). It is only used when intubation is not immediately possible. If pt is brought in by ambulance w/ the EOA in place, one must immediately consider removal & intubation. Prior to removing the EOA, the trachea must be intubated to protect the lungs from aspiration. The patient will usually vomit when the EOA is removed, so suction equipment must be ready. The EOA should not be used on those under age 16.

PMV   (PASSY-MUIR SPEAKING VALVE)

Description:
The PMV is a medical-grade plastic valve that allows patients to verbally communicate, improves secretion management, improves swallowing, reduces aspiration, & can expedite weaning. The PMV restores a more normalized physiology & breathing pattern, by restoring airflow through the oral & nasal cavities during exhalation. It is attached to the end of the tracheostomy tube. There are many different valves available depending on the patient's needs & application. Visit the website at: passy-muir.com

Utilization:
The PMV can be used on patients from infant to adult who have a trach tube. It can be used for spontaneously breathing & mechanically ventilated patients. The pt should be alert, responsive, & medically stable. The pt must be able to generate sufficient air during exhalation around the tube, & through the upper airway (may need many trial attempts to relearn normal breathing pattern). The PMV can be used during all waking hours when tolerance improves to that point.

The PMV should be removed for sleep. Monitor the patient's vital signs while the PMV is on.
Always deflate the cuff when the PMV is on, otherwise the patient will be unable to breathe.

Contraindications:
Unconscious pt. Extremely thick & copious secretions. Severe tracheal or laryngeal stenosis. Severe respiratory infection. Laryngectomy. **Do not use the PMV with a foam cuff tube, & do not use with an endotracheal tube.**

Cleaning Notes:
The PMV mfg outlines specific cleaning instructions: Clean daily by swishing in mild soap & warm water. Then rinse in water, & air dry. Do not use any other products or methods of cleaning.

SUCTIONING & VACUUM PRESSURES

Indications for suctioning:
Inability to cough effectively, inability to clear problematic secretions without assistance, & breath sounds with coarse rhonchi.
**Note: see appropriate suction catheter sizes in table above.

Vacuum Pressure – Acceptable Ranges:

Adults:	−80 to −120 mmHg
Children:	−80 to −100 mmHg
Infants:	−40 to −80 mmHg

The Suction Procedure:
Hyperoxygenate & hyperventilate the pt. Instill saline & insert suction catheter proper distance. Apply intermittent vacuum to aspirate secretions, keeping suction time less than 15 seconds. Oxygenate and ventilate patient as necessary. Assess patient for problems including bradycardia, skin color changes, decreased pulse ox saturation, & changes in respiratory rate. Observe for complications & hazards as mentioned below.

Hazards & Complications:
Watch for changes in vital signs. Bradycardia, arrhythmias, blood pressure changes, hypoxemia, laryngospasm, bronchospasm, uncontrolled coughing, gagging, vomiting, atelectasis, increased ICP, cardiopulmonary arrest. Most common: airway trauma & bleeding. Late complications are tracheal necrosis & respiratory infection.

Chapter 16

MECHANICAL VENTILATION, CPAP & BIPAP

ALSO SEE: *Artificial Airways & Suctioning – Chapter 15*
ABG's w/Formulas – Chapter 7
Hemodynamics w/Formulas – Chapter 8
PFT's – Chapter 13

Table 1. FORMULAS & NORMAL VALUES FOR MECHANICAL VENTILATION

Parameter	Abbrev	Formula	Normal Values
Airway Resistance	RAW	PPeak – PPlateau / flowrate L/sec	0.5 to 2.5 cmH_2O/L/sec
Airway Resistance Trend	RAW Trend	PPeak – PPlateau	<$10cmH_2O$ on vent. Keep RAW as low as possible.
Alveolar Ventilation	VA	VA = (VT – VD) x f	VA should closely parallel VE
Compliance	...	Compliance = ΔVolume / ΔPressure	...
Compliance, Dynamic	CD	(VT – tubing expansion volume) / (PIP – PEEP)	Varies based on CS & RAW
Compliance, Static (lung)	CS	(VT – tubing expansion volume) / (Plateau pressure – PEEP)	.1 to .2 L/cmH_2O or 100 - 200 mL/cmH_2O
Compliance Lung & Thorax	CLT	...	0.1L/cmH_2O
Cycle Time	Tc	(TI + TE) or (60 / f)	Varies based on insp time & rate
Deadspace Tidal Vol Ratio	VD/VT	($PaCO_2$ – $PECO_2$) / $PaCO_2$	20-40% off vent 40-55% on vent Anatomical VD = 1 mL/pound
Deadspace Mechanical	VD Mech	10 mL/inch	Near zero unless clinically indicated
Elastance	...	1/compliance	...
Expiratory Time	TE	(60 / f) – TI or Tc – TI	Varies based on insp time & rate
Flowrate	Flowrate	VE x (I + E) or VE / %TI decimal or VT / (TI / 60)	Set above pt demand; often 30-50 L/m, but can exceed 80 L/m.
Insp to Exp Ratio	I:E Ratio	TI:TE	1:2, 1:3, or 1:4

Inspiratory Time	TI	$(60 \div f) - TE$ or $Tc - TE$	0.8 to 1.2 seconds
Inspiratory Time %	TI%	TI / Tc	20% to 33%
Mean Airway Pressure	MAP or PMean	Most ventilation factors affect & ↑ or ↓ PMean. (i.e. +1 PEEP =+1 mean)	Keep as low as possible
Minute Ventilation	VE	VT x f	4-8 L/m off vent 5-10 L/m on vent
Negative Inspiratory Force	NIF	Measured w/ manometer. Also called MIP (maximal inspiratory pressure)	> 80 cmH$_2$O
Peak Inspiratory Pressure	PPeak or PIP	PPlateau / RAW or PIP can be viewed on manometer	< 40 cmH$_2$O. Keep as low as possible.
Plateau Pressure	PPlateau	PPeak – RAW (use inspiratory hold; can be viewed on manometer)	< 32 cmH$_2$O
Rate or frequency	f	$60 \div Tc$ or VE $\div$ VT	8-20 breaths/min spont. See Table 2 for setting ventilator breaths.
Resistance	...	Δ in Pressure / Flow	...
Shunt %	QS/QT	(A-aDO$_2$) x .003 / (CaO$_2$ -CvO$_2$) + (A-aDO$_2$) x .003	Normal 5% Moderate defect 10-20% Severe defect > 20%
Tidal Volume	VT	VE / f	7 mL/kg spont breaths 6-10 mL/kg vent breaths
Tubing Expansion	Tubing Exp Vol	Adult circuit: 3 Neonate circuit: 1	Tubing Expansion Volume is in mL/cmH$_2$O

Notes: 1 mL = 1 cc

1 cmH$_2$O = 1.36 torr or mmHg

Notes: This chapter contains adult values provided as a reference only. Clinicians should also refer to the policies & protocols set forth by the employer.

INDICATIONS FOR MECHANICAL VENTILATION

- Acute Respiratory Failure (ARF)– Hypercapnic or Hypoxemic
- Impending Respiratory Failure
- Apnea
- Inability to correct ARF by other methods

125

Table 2. SETTINGS FOR MECHANICAL VENTILATION
(These are initial generic settings for adults that can be utilized
when little information is available about the patient)

Rate.................10-15 / minute
VT......................6-10 mL/Kg (Previously 8-12 mL/Kg was recommend)
VE....................5-10 L/m
O_2.......................40-60%*
PEEP.................+5 cmH_2O**
PS.....................+5 to +10 cmH_2O
Flowrate.............30-50 L/m***
I:E Ratio............ 1:2, 1:3, or 1:4
Inspiratory time.. 20% to 33% or 0.8 seconds to 1.2 seconds
Sensitivity.......... – 0.5 to –2 cmH_2O or
                 flow trigger 1-3 L/m below baseline or bias flow.

*Initial O_2 up to 100% depending on the situation (i.e. cardiac/
respiratory arrest, poor pulse ox saturation, cyanosis, etc.)
**PEEP can be set to zero in cases of low blood pressure.
***Must meet or exceed pt demand; Pt demand can exceed 80 L/m.

Notes: As clinical data becomes available & patient response to
ventilation is observed, the ventilator settings are adjusted (see Table
3). Alarms must also be set appropriately (discussed at end of chapter).

Table 3. ADJUSTING SETTINGS ON MECHANICAL VENTILATION
CONTINUOUS PATIENT ASSESSMENT
& WEANING PARAMETERS

Parameter	Normal Values	Acceptable Range on Ventilator & Acceptable Range for Weaning
Respiratory rate (spontaneous)	8-20 / minute	8-20 / minute
Tidal volume (spont)	7 mL/kg	> 5 mL/kg
Minute ventilation (spontaneous)	5-10 L/m	5-10 L/m
Vital capacity	65 mL/kg	> 10 mL/kg
NIF	– 80 cmH2O	– 20 cmH_2O or better
A-a Gradient	25-65 mmHg on 100% 10-15 mmHg on 21%	< 300 mmHg on 100%
QS/QT	< 5%	< 20%

VD/VT	40-55% on Vent 20-40% off Vent	<60%
Pulse Oximetry SpO$_2$	> 95%	> 95%
ABG's..............pH .... PaCO$_2$ ....PaO$_2$	7.40 40 mmHg 100 mmHg	7.35 - 7.45 35 - 45 mmHg > 70 mmHg

Additional parameters to monitor:

Monitor the Patient:
Overall medical condition, vital signs, skin color, nutritional status.
EKG, hemodynamics, laboratory tests, x-rays.
Breath sounds, secretions, & work of breathing.
Provide suctioning and bronchodilator as necessary.
Is the patient progressing & meeting goals on current settings?
Does the patient need parameter changes?

Monitor the Equipment:
Check for proper function of the ventilator.
Assess all parameters, pressures, volumes & alarms.
Assure the equipment is clean & problem free from the
        ET tube or trach tube to the ventilator.
Be sure the circuit is clear of excess condensate.
Provide adequate humidity w/ an HME or heated humidifier.
Secure the ET/trach tube & utilize proper cuff pressure.
Elevate the head of the bed.
Keep a resuscitation bag & all emergency equipment on hand.
Change ventilator circuits, filters, & suction catheters regularly.
Perform equipment calibration, cleaning, & maintenance regularly.
(Also see the Troubleshooting section of this chapter)

***** MODES OF VENTILATION *****

Note: For each of these modes, be certain to set all applicable parameters appropriately (see samples in Table 2). Also assure all alarms are set appropriately. Modes are listed alphabetically.

AC or VC – Assist Control or Volume Control

AC or VC provides full ventilatory support. Every breath is delivered at preset VT (time cycled & pt triggered breaths are given w/ the same VT). A guaranteed minimum VT, rate, & minute volume is provided. The pt can trigger additional spontaneous breaths as long as sensitivity is set appropriately. The spont breaths will be given at the same VT as vent breaths (PS is not an option). AC or VC is a good choice for pt's with ↓ sensorium who need full support, as long as lung mechanics are normal & peak pressures remain low. However, consider SIMV as an alternative, as SIMV is better tolerated, & AC causes higher mean pressure & hyperventilation if spont rate rises.

APRV – Airway Pressure Release Ventilation

APRV is designed to recruit collapsed alveoli & optimize ventilation while minimizing barotrauma. Useful in treating ARDS & ALI. In simple terms, APRV is described as CPAP w/ regular, brief, intermittent releases in airway pressure. Release phase facilitates CO_2 removal & can be pt triggered. APRV uses long inspiratory time to improve oxygenation. APRV maintains patency of alveoli by maintaining a constant pressure w/ fixed inspiratory time, & tidal volumes are delivered during transient decreases in intrathoracic pressure. In clinical trials, APRV was better tolerated than IRV, gave lower PIP's & gave similar improvements in oxygenation.

AUTOMODE

Automode is a combination of modes that automatically facilitates weaning at the earliest possible opportunity, as soon as the pt triggers 2 spont breaths. The vent mode changes back to full support if there is an apneic period. This mode may not be appropriate for those who need VT, rate, & VE closely monitored & controlled. It is important to set all applicable parameters & alarms for both applicable modes. Combination modes available are: (PRVC & VS), (VC & VS), or (PC & PS). For more information on these modes, please refer to each mode as listed separately in this section.

CPAP – see PEEP

ECMO – Extracorporeal Membrane Oxygenation

ECMO is effective for treating newborns w/ acute respiratory failure, & is also used for treating severe acute hypoxemic respiratory failure, including ARDS. ECMO involves an arteriovenous circuit for diverting the majority of cardiac output through an artificial lung that facilitates exchange of CO_2 & O_2.

HFV & HFO – High Frequency Ventilation & Oscillation (for adults)

HFV is a mode of ventilation utilizing rates over 60. Terms used are HFV, HFJV, HFPPV & HFO. All of these are referred to as "HFV", as in HF Ventilation, HFJet Ventilation, HF Positive Pressure Ventilation & HF Oscillation or HF Oscillatory Ventilation. The term hertz or Hz means cycles per second. **1 Hz = rate of 60**. One can simply multiply Hz by 60 to figure out the rate (i.e. 4 Hz is a rate of 240.) HFO or HFOV is a term that technically means the rate is 300 or higher (5 Hz or higher). <u>In HFV-HFO the rates are very high at 61 to 2000 or greater. The tidal volumes are very low at about 2 to 5 mL/kg.</u> The tidal volumes can be slightly above deadspace ventilation, equal to, or even lower than deadspace ventilation. HFV is designed to provide adequate ventilation while minimizing alveolar collapse via high **expiratory** lung volumes. HFV has been utilized for decades for neonates & proven effective. HFV is currently being utilized for adults in cases like ARDS, to improve oxygenation while minimizing barotrauma, when other modes of ventilation fail. The pt must be sedated when utilizing HFV & HFO.

IRV – Inverse Ratio Ventilation

IRV is utilized to recruit alveoli & improve oxygenation w/ prolonged inspiratory time. IRV is used to treat ARDS & other cases when other methods of improving oxygenation fail. The I:E ratio usually ranges from 1.5:1 to 4:1. The patient will need to be sedated & sometimes paralyzed, as IRV is uncomfortable to the pt. Some success has been reported in studies where IRV is used to improve oxygenation in ARDS patients.

IVOX – Intravascular Membrane Oxygenation & carbon dioxide removal

IVOX devices are currently undergoing trials for use in lung protective ventilatory strategies in cases like severe respiratory failure. IVOX has proven to remove up to 30% of CO_2 in patients w/ near normal CO_2 levels, thereby allowing reduction of IPPV.

Liquid Perfluorocarbon Mechanical Ventilation

Liquid ventilation involves perfluorocarbon-associated gas exchange. This mode is currently being studied to treat ARDS. Perfluorocarbons dissolve O_2 & CO_2, thereby facilitating pulmonary gas exchange. The lungs are partially filled w/ the fluid, & mechanical ventilation is provided using conventional ventilators.

Negative Pressure Ventilators (Extrathoracic)
Includes the Iron Lung & Chest Cuirass

Negative pressure, by way or suction, is applied to the outside of the chest. This causes inspiration by making the chest to rise & expand. Ventilation is controlled by adjusting the amount of negative pressure & the inspiratory time. These ventilators are not used commonly, as modern-day ventilators utilize positive pressure. However, these ventilators are still utilized for long-term management of some chronic respiratory failure patients.

PC - Pressure Control

PC mode provides full ventilatory support. A minimum guaranteed rate is dialed in, w/ preset pressures, w/ varied tidal volumes. Breaths are time cycled & also can be pt triggered. The inspiratory phase is pressure limited, & delivered at preset PC above PEEP level. Flowrate is high initially to meet pt demand, & then decelerates. PC is a good choice for pt's w/ unacceptable high pressures in other modes such as SIMV or AC. ARDS is one case where PC may help avoid barotrauma. PC can be used with IRV if the pt needs prolonged insp time. One must be alert to significant changes in VT & VE since those can change to unacceptable levels.

PEEP / CPAP - Positive End-Expiratory Pressure
or Continuous Positive Airway Pressure
(Also see CPAP & BiPAP stand-alone devices at end of this section)

PEEP/CPAP describes breathing at greater than zero baseline pressure. PEEP/CPAP optimally will improve oxygenation, increase compliance, increase FRC & decrease atelectasis. One should add or increase PEEP if the PaO2 is < 60 on FiO2 > 40%. The common range of PEEP used is +3 to + 12 cmH$_2$O, and higher if indicated. PEEP of + 3 to + 5 cmH$_2$O restores physiologic PEEP lost due to an artificial airway. Use PEEP cautiously in those with low blood pressure. Also use cautiously in those who with increased FRC. PEEP/CPAP is often combined w/ other modes when utilized on conventional ventilation. Technically, the terms is PEEP when there is a mechanical rate, & CPAP when all breaths are spontaneous. Although PEEP & CPAP are commonly used interchangeably. (Also see Complications section & Auto-PEEP in the troubleshooting section).

Optimal PEEP is the least amount of PEEP necessary to provide adequate oxygenation without adverse effects. Parameters that verify Optimal PEEP are:

- PaO$_2$ is increasing.
- Static compliance is increasing.
- Hemodynamic pressures are remaining stable.

130

- Blood pressure & cardiac output are **not** unacceptably decreasing.
- PAP & PWP are stable, **not** unacceptably increasing.

PRVC - Pressure Regulated Volume Control

PRVC provides full ventilatory support. A guaranteed minimum rate is dialed in. Vent breaths & spontaneous breaths are delivered at preset VT. PRVC is a pressure limited mode w/ preset inspiratory time. The flowrate is initially high to meet pt demand, then decelerates. The main difference between VC & PRVC is: in PRVC the vent automatically adjusts the inspiratory pressure level based on changes in lung mechanics on each breath. Changes occur in increments of 1 to 3 cmH_2O. PRVC always uses the lowest pressure needed to deliver the desired VT & VE. PRVC is a good choice for pt's with little or no spont breathing capacity, and have unacceptably high pressure in other modes. It's important to set upper pressure limit appropriately since upper pressure limit serves 2 purposes in this mode; 1st it is an alarm, & will automatically switch to exhalation when reached, so full VT will **not** be delivered. 2nd it should be set as low as possible while ensuring adequate tidal volume. Whenever the PIP reaches a point 5 cmH_2O below upper pressure limit, a "Limited Pressure" alarm is activated & the delivered VT may be less than preset VT. Since VT varies, one must be alert to any significant changes in VT & VE.

PS or PSV - Pressure Support Ventilation

PS mode is for pt's who are spontaneously breathing, but still need partial ventilatory support. PS is not to be used for those w/ little or no respiratory drive. PS assists each spontaneous breath w/ inspiratory pressure at the dialed-in level. PS helps to increase the spont VT, thereby improving ventilation & oxygenation. PS also helps decrease WOB. PS can be used alone, with PEEP/CPAP, or combined w/ certain modes. Insp flowrate is high initially to meet pt demand, then decelerates. PS +5 helps overcome resistance of breathing through the circuit. A commonly used range of PS is +5 to +15 cmH_2O, but +20 or higher is used PRN. PS is often set to achieve a target VT of 6 to 10 mL/kg.

SIMV / IMV - Synchronous Intermittent Mandatory Ventilation

SIMV can provide full or partial ventilatory support. SIMV + PS is a good choice for most patients. SIMV can be utilized for a wide range of support from apnea, to some spont breathing, to weaning & extubation. SIMV is not a good choice for pt's with consistently high & unmanageable PIP. In SIMV the ventilator provides a guaranteed minimum VT, rate & minute volume. The dialed-in rate is provided at

131

the preset VT. Additional spontaneous breaths can be purely spontaneous, or more commonly, supported w/ PS.

SIMV PC+PS - SIMV Pressure Control + Pressure Support

SIMV PC+PS can provide full or partial ventilatory support. This mode combines PC & PS with a synchronized rate. (Please refer to PC & PS listed individually for further info.) Here the vent delivers the set rate of mandatory breaths using the PC mode, & assists w/spont breaths using the PS mode. The VT varies, so one must be alert to significant changes in VT & VE. Insp flowrate is high initially to meet pt demand, then decelerates. This mode is indicated for those who have unacceptably high PIP in other modes, those needing high flowrates & have some spont breathing. Special notes about this mode: 1st It is possible for the pt to have some purely spont breaths. 2nd One common problem causing pt resp distress is a mismatch between PC & PS breaths. So it is often best to begin w/ PC & PS breaths **at or near** the same pressure. (i.e. set both at +20)

VS - Volume Support

VS mode is for spontaneously breathing patients who still require partial ventilatory support. VS is also utilized as a weaning mode, w/ preset backup in case of apnea. VS is not indicated for those w/o adequate respiratory drive. VS helps decrease WOB, increase ventilation, & improve oxygenation in spontaneously breathing pts. VS differs from PS in that VS provides automatic weaning of PS as long as the pt's tidal volume matches the minimum VT required. Insp flow is decelerating, giving high initial flowrate to meet pt demand. Each breath is pt triggered, & the vent uses the lowest PS necessary to achieve preset minimum expected VT & VE. The vent automatically adjusts insp PS in increments of 1 to 3 cmH$_2$O. The desired rate is dialed in & is part of the automatic vent calculations; but, the actual rate & insp time are pt dependent, unless the pt has an apneic period. The automatic backup is PRVC, w/ same set VT & set backup rate. Target VT should be set to achieve a target VT of 6-10 mL/kg. Set all applicable parameters listed in Table 2 & alarms appropriately. Important to set upper pressure limit appropriately since upper pressure limit serves 2 purposes in this mode; 1st it's an alarm, & will automatically switch to exhalation when reached; thus full VT won't be delivered. 2nd it should be set as low as possible while assuring adequate tidal volume. Whenever the PIP reaches a point 5 cmH$_2$O below upper pressure limit, a "Limited Pressure" alarm is activated & the delivered VT may be less than preset VT. (VT can also ↑; the ventilator will deliver up to 150% of the set minimum VT). Since VT varies, the clinician must be alert to significant changes in VT & VE.

132

CPAP & BIPAP THERAPY*
(Stand-alone devices)

This section regards mainly stand-alone CPAP & BIPAP machines, & the controls found on these machines. (Also for more information, refer to PEEP & PS above, in the Modes of Ventilation section.) Precautions & hazards of CPAP & BIPAP are the same as those for PEEP & PS. One should avoid excessive airway pressures, monitor blood pressure (especially if using high pressures), monitor vital signs, & set alarms accordingly. The incidence of pt discomfort & non-compliance can be minimized by: Assuring correctly fitted mask/nasal device to maintain desired pressure & improve pt comfort. Thoroughly educate pt prior to starting therapy.

CPAP
Description: Continuous Positive Airway Pressure (CPAP) is the same therapy as PEEP, but the term CPAP is used when the patient is spontaneously breathing, w/ no mechanical ventilation rate. CPAP is spontaneous breathing at pressures above zero. One normally sets CPAP at +3 to +12 cmH_2O, or higher PRN. CPAP increases FRC, improves compliance, decreases atelectasis, & improves oxygenation. Also used for obstructive sleep apnea (OSA) to overcome upper airway obstruction by splinting the airway open.
Ramp: The ramp button, when pressed, causes a decrease in CPAP pressure, that slowly builds back up to the set pressure. Many patients find it easier to start w/ the low pressure & slowly get used to the higher pressure, or may fall asleep before the machine reaches set pressure.
C-Flex: C-Flex is a very brief drop in pressure at the beginning of **exhalation**, after which pressure quickly returns to set pressure. Designed to improve pt comfort, C-flex makes it easier to exhale. The pressure drop is variable based on pt need & tracked by the machine on a breath-by-breath basis. C-Flex can be set between 1 and 3, w/ 1 being the least pressure relief, & 3 being the highest pressure release.
Auto-CPAP / Smart CPAP: Automatically adjusts pressure to meet pt needs, & always uses the lowest pressure necessary. The clinician programs minimum & maximum CPAP levels into the machine. Used mainly for OSA patients when further study of best pressure is needed, or for those w/ challenging breathing patterns during sleep. A "smart card" computer disk can be inserted in the machines to record sleep events & help track optimal pressures.

BIPAP

Description: Biphasic (or bilevel) Positive Airway Pressure (BiPAP) is similar to a combination of CPAP & Pressure Support. This is also called NPPV or non-invasive positive pressure ventilation. BiPAP is useful in providing ventilatory support for patients w/ respiratory insufficiency. Most machines come equipped w/ alarms, & may have option of adding a mechanical backup rate. Newer machines provide display screens w/ multiple options, & can also be used w/ trach pts.

IPAP: Inspiratory Positive Airway Pressure is similar to pressure support. IPAP improves ventilation by increasing VT, decreasing WOB & thus improves oxygenation. Set IPAP above EPAP level, usually between +5 and +20 cmH$_2$O, but can be higher PRN.

EPAP: Expiratory Positive Airway Pressure. Similar to PEEP/CPAP. EPAP is spontaneous breathing at pressures above zero. Usually set at +3 to +12 cmH$_2$O, or higher PRN. Increases FRC, increases compliance, decreases atelectasis, & improves oxygenation. Also used for OSA to overcome upper airway obstruction as the pressure serves to splint the airway open.

DESCRIPTIONS OF OTHER VENTILATOR CONTROLS

Expiratory Retard
  Prolongs exhalation & decreases the FRC, which is the opposite effect of PEEP. Mimics pursed-lip breathing. Useful w/ COPD pts.

Flowrate
  Peak inspiratory flow of gas. Must be adjusted (depending on mode) to meet or exceed patient demand. Usual range is 30-50 L/m, but can exceed 80 L/m. If set too high, can increase RAW & PIP. If set too low, can cause increased WOB, and make TI unacceptably long.

I:E ratio
  The normal ratio is 1:2, 1:3, or 1:4. COPD patients need more time to exhale, therefore a longer I:E ratio, 1:3 or 1:4. Ratio of 1:1 and inverse ratios are also utilized when clinically indicated.

Inspiratory Pause or Plateau
  An inflation hold maneuver where the patient is made to hold the tidal volume at the end of inspiration for a short time. If used, it's usually set at 0.5 to 1 second. Improves distribution of ventilation & can increase PaO$_2$, increase inspiratory time & decrease expiratory time, thereby decreasing the I:E ratio. If there is a need to maintain I:E ratio & add the inspiratory pause, one can increase the flowrate to decrease the inspiratory time.

Inspiratory time
  Important time parameter where inspiratory time is set by clinician, then the ventilator calculates the remaining time as expiratory time, as calculated by the preset rate. This also sets the I:E ratio. Usual range

for TI time is 0.8 to 1.2 seconds, or 20% to 33%. Too short TI along w/ low flowrate can result in decreased VT. Too long TI can cause pt discomfort & too short TE.

Pressure Limit
A peak pressure alarm & also a parameter to set in certain modes (as described in applicable modes). The upper pressure alarm is usually set about 10 above actual or desired PIP.

Pressure Rise Time or Slope
Used to slow the rate at which peak inspiratory pressure is reached. If used, assure inspiratory gas flow meets or exceeds pt demand. If too slow, WOB can increase.

Sensitivity
Used to allow pt to trigger breaths. Usual range is − 0.5 to −2 cmH$_2$O, or flow trigger 1-3 L/m below baseline or bias flow. Must be sensitive enough for pt to easily trigger breaths without increasing WOB, but assure it's not so sensitive that ventilator auto-cycles.

Sigh
Can be used to simulate normal breathing sighs, but not often used of late. Helps prevent atelectasis. If used, set at up to 10 per hour at up to 1 ½ times the set tidal volume. Monitor PIP to assure it is not too high during sighs. Sigh will increase VT, PIP & mean pressure.

Waveforms / Flow Patterns
Square wave flow pattern: Produces a constant flow throughout inspiration. Useful for non-compliant lung conditions including ARDS and pneumonia.

Sine wave flow pattern: Simulates normal breathing pattern. Flow begins slowly & increases to a peak, then decreases at the end of inspiration. Provides good distribution of ventilation.

Accelerating flow pattern: Flow begins slowly & rises to a peak, then ends at the end of inspiration. Can reduce turbulent flow & decrease mean airway pressure in those with high RAW.

Decelerating flow pattern: Flow is very fast during the beginning of inspiration, peaks, then gradually decreases. Very useful for patients who need high flow rates.

VENTILATOR ALARMS

Alarms available on most ventilators include high & low pressure, high & low rate, and high & low minute volume. These alarms should be set near the patient's current actual parameters, so alarms will activate w/ any significant changes. Also adjust alarm volume so it is clearly audible. Apnea alarm should be set w/ minimal delay time for onset of alarm. If backup ventilation is available, set all ventilating parameters for both modes accordingly for the patient.

COMPLICATIONS OF
MECHANICAL VENTILATION & PEEP/CPAP

The most common complications associated w/ mechanical ventilation are due to positive airway pressure. Higher pressures increase risk of complications. The complications of Positive Pressure Ventilation (PPV), Intermittent Positive Pressure Ventilation (IPPV) & PEEP/CPAP include:

Barotrauma, pneumothorax, pneumomediastinum, SQ emphysema, ↑ PVR, ↑ PAP, ↓ venous return, ↓ cardiac output, ↓ blood pressure, edema due to ↑ antidiuretic hormone (ADH) & ↓ urine output.

Inappropriate settings on machine can result in hyperventilation or hypoventilation. There is also the potential hazard of machine malfunction. Also see hazards of artificial airways, intubation, & suctioning in Chapter 15.

TROUBLESHOOTING
& ADJUSTING PARAMETERS

Problem: Abnormal ABG's
Solutions: Most common methods of correcting abnormal ABG's:

Adjust PaO_2 – To increase PaO_2: ↑ FiO2, ↑ PEEP, ↑ Rate, ↑ VT
– To decrease PaO_2: ↓ FiO2, ↓ PEEP, ↓ Rate
FiO_2 needed = (desired PaO_2 x current FiO_2) / Current PaO_2
(If >60% FiO_2 required, consider increasing PEEP)

Adjust $PaCO_2$ – To increase $PaCO_2$: ↓ Rate, ↓ VT, ↓ VE
– To decrease $PaCO_2$: ↑ Rate, ↑ VT, ↑ VE
Desired VE = current VE x (current $PaCO_2$ / desired $PaCO_2$)

Problem: Low pressures or low exhaled volumes
Solutions: Find & correct the leak. Most likely in the cuff, the circuit, or attachments.

Problem: Increased spontaneous respiratory rate
Solution: Treat the cause. Most common causes include hypoxemia, acidosis, pain, or anxiety.

Problem: High peak pressures
Solutions: If problem is due to ↑ RAW: Suction secretions, give bronchodilator, ↓ flowrate, assure condensate is drained, assure no tubing kinks, assure ET or Trach tube is not malpositioned. Check if pt is biting tube & add bite block if necessary. Also Δ waveform, remove any insp pause or hold. Can consider ↓ VT (also need to ↑ rate to maintain VE).
If problem is due to ↓ lung compliance: Treat the cause. For pneumothorax- chest tube. For pneumonia- antibiotics, CPT, etc. For ARDS- ↑ PEEP, ↑ TI, Δ mode to PC.

Problem: Auto-PEEP or Inadvertent PEEP
Can be seen by performing a brief exp pause, & watching the increase in pressure on the manometer.
Solutions: To decrease auto-PEEP: Decrease airflow obstruction by providing adequate bronchodilator therapy & suctioning the airway. Also ↓ VT, ↓ TI, ↑TE, ↑ flow rate, & use square wave flow pattern.

Problem: Inspiratory time too long/ I:E ratio too high
Solutions: ↓ inspiratory time, ↑ flow rate, ↓ VT, remove any inspiratory hold or plateau.

Problem: Patient fighting the ventilator
Solutions: Check that all parameters are set appropriately for the patient, & readjust as necessary. Patient may need higher flowrate, or adjusted sensitivity. If ↑ PIP, see high pressure problem. Check vital signs. Evaluate if patient needs sedation, or is ready for weaning.

Problem: Ventilator failure to cycle or other machine malfunction
Solutions: Check for appropriate sensitivity setting. If that is not the problem, remove patient from ventilator. Have one therapist provide manual ventilation & oxygen w/ resuscitation bag. Have another therapist check ventilator functions & also check for inadvertent changes made to vent settings; correct problem or replace ventilator.

Problem: Increased intracranial pressure (ICP)
Solution: Using Hyperventilation to Decrease ICP
Hyperventilation can be used to decrease intracranial pressure (ICP) if the ICP rises above 10. The normal ICP value is 0-15 for adults and 0-10 for children. Patients w/ head trauma, intracranial swelling, or other conditions that raise ICP can benefit from deliberate hyperventilation to a mild state of respiratory alkalosis. Normally this is done by increasing the rate above normal, & maintaining normal tidal volume. Hyperventilation causes cerebral vasoconstriction, resulting in decreased cerebral blood flow, & decreased ICP. This is a temporary measure, since the kidneys will compensate after a few days.

Chapter 17

NEONATAL/PEDIATRIC RESPIRATORY CARE

Fetal Pulmonary Data

Weeks 0 to 5 – The Embryonic Period
Weeks 1 to 6 – The Pseudoglandular Period
Weeks 17 to 24 – The Canalicular Period
Weeks 24 to 40 – The Terminal Sac Period

- At 24 weeks gestation, the respiratory bronchioles, alveolar ducts, alveoli, & Type II cells develop.
- At 30-32 weeks gestation, the alveoli continue to develop, along with alveolar ducts, respiratory bronchioles & capillaries. The surface area grows considerably. At around 32 week's, the lecithin/sphingomyelin ration (L/S ratio) changes. Prior to 32 weeks, there is more sphingomyelin than lecithin present; making the L/S ratio less than 1. Then around 32 weeks, lecithin production rapidly increases and sphingomyelin production decreases.
- At about 35 week's gestation, the L/S ratio reaches 2. The phospocholine transferase system develops, which is a key pathway for lecithin production. An L/S ratio less than 2 means the lungs are not well developed & the possibility of hyaline membrane disease is increased.
- At 40 weeks gestation (full term), the L/S ratio reaches between 4 and 6.

Approximately 33% fetal lung fluid exits the lungs before labor begins. During labor, the vaginal squeezes help to expel about 20%. The remainder of fetal lung fluid is expelled within the first 24-hour period after the birth of the infant.

Fetal Blood Gases: (normal values)
Umbilical Vein: pO_2 is about 30 torr & about 80% saturation.
                    pCO_2 is about 40 torr.
                    Bicarb is about 18 mEq/L.
Deoxygenated blood exits the fetus & then returns to the placenta through two umbilical arteries. During this period, oxygen diffuses into the fetal blood. This oxygenated blood then returns to the fetus through the umbilical vein.

Neonatal Data

Changes in the Pulmonary System at Birth:
At birth, the fetal lungs still contain considerable fluid. The infant must overcome the surface tension, forces of tissue resistance, and viscosity of fluid in the airways. This requires about -40 to -80 cmH_2O on the first inspiration.

Neonatal Arterial Blood Gases: (normal values just after birth)
        pO_2 is rises from 30 torr to 60 torr & about 90% saturation.
         pCO_2 decreases from about 55 to 35 torr over the first 3 hours.
         Bicarb decreases from about 23 to 20 mEq/L, where it
            remains for several days, & slowly rises to normal.

Assessing the Neonatal Patient:

Apgar Scores:

	0	1	2
Heart Rate	absent	under 100/minute	over 100/minute
Respiratory Effort	absent	slow, irregular	active crying
Muscle Tone	limp	slight flexion of extremities	active
Reflex Irritability	no response	grimace	coughing or sneezing
Color	blue or pale	blue extremities & pink body	pink throughout

Apgar score range is 0 to 10.
Apgar score at 1 minute: (In all cases, maintain thermal environment.)
   Score: 7 to 10 is normal. Monitor patient.
   Score: 4 to 6 indicates a moderately depressed neonate. Provide tactile stimulation w/slapping of the feet or drying w/ towel. Provide oxygen w/bag & mask using 5 lpm or less. Provide manual ventilation PRN. Watch for improvement, & if none, treat as score of 0 to 3.
   Score: 0 to 3 indicates a severely depressed neonate requiring emergent care. Suction the airway, insert ET tube, ventilate, & perform chest compressions at 100-120/minute. Also provide drugs including epinephrine, bicarb, & others (See Chapter 4 RT Medications).

Important Notes Regarding Newborns:
- **The airways must be cleared. This may require bulb suction to mouth & nares or deep suction if required.**
- **The infant must be dried and provided with stimulation by drying with towel. (One can be more vigorous with stimulation depending on the Apgar score.)**
- **Thermoregulation is important. Keep the infant warm & maintain the infant's environment utilizing blankets, caps radiant warmer, or incubator as needed.**

Breathing Patterns in Neonates:
- Normal respiratory rate for a full-term infant is 30-60 per minute.
- Normal tidal volume averages 6-8 mL/kg.
- Normal heart rate is 120 to 130 per minute.
- Periodic breathing is often seen in premature infants. Apneic episodes of 5 – 10 seconds, with periods of ventilation lasting 10 – 15 seconds.
- Apneic periods lasting over 20 seconds can result in; respiratory acidosis, decreased 02 saturation, and bradycardia. Providing oxygen and CPAP are methods that may correct the problem.

Infant & Pediatric Normal Values
Respiratory Rate, Heart Rate, Blood Pressure

Age	Respiratory Rate	Heart Rate	Blood Pressure
Newborn	30 – 40	125	72/48
1 Year old	25-35	120	90/60
2 Year old	25-35	110	90/60
3 Year old	20-35	105	95/60
4 Year old	20-35	100	100/60
7 Year old	15-25	90	110/60
14 Year old	15-20	80	120/70

*Note: Pre-term neonates with below average birthweight usually have lower blood pressures. I.E. at about 2000 grams, the blood pressure is about 50/30.

Neonatal & Pediatric Artificial Airways & Suction

Note: Refer to Chap 15 Artificial Airways & Suction for more information.

Age	ET Tube Size Internal Diameter (mm)	Laryngo- scope Blade Size	Suction Catheter (French size)
Newborn < 1000 grams	2.5	0	5
1000-2000 grams	3.0	0	6
2000-3000 grams	3.5	1	6-8
> 3000 grams	3.5 – 4.0	1	6-8
6 months	3.5 – 4.0	1-2	6-8
1 year	4.0	1-2	8
2-3 years	4.5 – 5.0	2	8
4-7 years	5.0 – 6.0	2	8-10
8-10 years	6.0 – 6.5	2-3	10
12 years	7.0	2-3	10
15 years	7.0 – 7.5	3	10-12

Cuff Pressures

Many neonatal size tubes are cuffless. If a cuff is present, the Tube Cuff pressure should always remain below 20 mmHg (26 cmH$_2$0) to prevent edema & necrosis.

Vacuum Pressure – Acceptable Ranges:

Children: –80 to –100 mmHg
Infants: –40 to – 80 mmHg

Neonatal & Pediatric Oxygen & Ventilation

Note: Excessive oxygen utilization and high PaO$_2$ (>80) should be avoided whenever possible in infants, especially if premature. Problems include retinopathy of prematurity (ROP), retrolental fibroplasia (RLF), & broncho-pulmonary dysplasia (BPD). Also, remember that thermoregulation is very important for newborns.

Oxygen Delivery Device	Oxygen Flow Rate LPM	FiO$_2$ % (appx)	Notes
Nasal Cannula Neonatal	¼ – 2	22% to 45% or higher PRN	For Neonatal NC, the source gas is usually not at 100%. Source gas is reduced by air entrainment via blender to achieve desired FiO$_2$.
Nasal Cannula Pediatric	1 – 6	24% to 44%	Appx values using 100% O$_2$ source gas: 1 lpm=24%, 2 lpm= 28%, 3 lpm= 32%, 4 lpm= 36%, 5 lpm= 40%, 6 lpm= 44%
Oxygen Hood	7 – 12	21% to 100%	Keep flow 7 or higher to flush out CO$_2$. Maintain thermal environment at 32° - 34°C. Gives precise FiO$_2$. Good access to patient.
Oxygen Tents: Croup Tent, Mist Tent, Aerosol Tent	10 – 12	21% to 40%	May be utilized for children needing O2, but unwilling to use mask or NC. FiO$_2$ is not precise. Fire hazards: don't use alcohol or oils on pt, & don't allow toys that may spark.

Other O$_2$ Delivery Devices- see Chapter 5.

Therapeutic Modalities – see Chapter 5.

CPAP for Infants:

Indicated for infants with PaO_2 <50 on FiO_2 >=50%, with acceptable $PaCO_2$. Useful for newborns with increased WOB, tachypnea, grunting, nasal flaring, apnea of prematurity & other disorders as listed in this chapter.

Initial CPAP: 4-5 cm H_2O & increase in +1 to +2 increments to +10 or higher PRN. Monitor for increased oxygenation, increased SpO_2, & decreased WOB, w/problems resolved before weaning. Wean oxygen to 40%, then wean CPAP off and place on other O2 delivery device with slightly higher FiO_2.

If CPAP fails, mechanical ventilation may be indicated. Failure recognized by: being unable to improve PaO_2, unable to decrease WOB, & $PaCO_2$ is rising.
Note: See Chapter 16 for more info on CPAP, hazards, complications, etc.

Mechanical Ventilation

Please refer to Chapter 15 for info on intubation, extubation, suctioning, artificial airways, etc.
Please refer to Chapter 16 for more information on many modes of ventilation, along with indications, setup, parameters, monitoring, hazards, complications, weaning, etc.

Pressure Ventilation

For infants & young children, positive **pressure** ventilation is normally utilized instead of volume ventilation. For initial ventilator setup, the parameters vary greatly based on the patient's needs. One must assess the patient, monitor the saturation, ABG's, and the patient's response to ventilation.

General guidelines for initial setup for an infant or young child utilizing pressure ventilation include:

	Infant	Young Child
Rate BPM	20 – 40	12 – 25
PIP cmH$_2$O	15 – 20	20 – 25
PEEP cmH$_2$O	2 – 6	2 – 10
FiO2 %	40-60 or higher PRN*	40-60 or higher PRN
VT cc/Kg	5 – 7	5 – 8
TI seconds	0.4 – 0.6 or up to 1.0 if indicated	0.5 – 1.0
TE seconds	0.5 – 1.5	0.5 – 1.5
I/E ratio	1:1 to 1:1.5	1:2

*Newborn: Titrate oxygen to keep SpO_2 88-93%. This helps avoid RLF & ROP. One month old & up (if not at risk for BPD), use standard protocol, generally titrating oxygen to maintain SpO_2 >92%.

HFV & HFO – High Frequency Ventilation & Oscillation

HFV is a mode of ventilation utilizing rates over 60. Terms used are HFV, HFJV, HFPPV & HFO. All of these are referred to as "HFV", as in HF Ventilation, HFJet Ventilation, HF Positive Pressure Ventilation & HF Oscillation or HF Oscillatory Ventilation. The term hertz or Hz means cycles per second. **1 Hz = rate of 60**. One can simply multiply Hz by 60 to figure out the rate (i.e. 4 Hz is a rate of 240). HFO or HFOV is a term that technically means the rate is 300 or higher (5 Hz or higher). In HFV-HFO, the rates are very high at 61 to 2000 or greater. The tidal volumes are very low at about 2 to 5 mL/kg. The tidal volumes can be slightly above deadspace ventilation, equal to, or even lower than deadspace ventilation. HFV is designed to provide adequate ventilation while minimizing alveolar collapse via high **expiratory** lung volumes. HFV has been utilized for decades for neonates & proven effective. HFV can improve oxygenation while minimizing barotrauma, when other modes of ventilation fail. The pt is sedated when utilizing HFV & HFO.

NOTES:

143

Neonatal & Pediatric Pulmonary & Cardiac Disorders

Many factors can contribute to high risk neonatal deliveries, including:

- Mother with: pre-eclampsia, eclampsia, hypertension or hypotension, diabetes, drug/alcohol abuse, prior or current C-section, premature rupture of membranes, age below 16 or above 35.
- Anesthesia and other drugs used during labor and delivery.
- Premature or post-term delivery.
- Abnormal fetal position.
- Trauma.

NOTATIONS:
S/S: **Signs & Symptoms**
Tests: **Diagnostic Tests** (emphasizing respiratory tests)
Tx: **Treatment**   (highlighting respiratory treatments)

Apnea of Prematurity
A condition of the immature newborn marked by repeated episodes of apnea. Usually resolves over time. Thought to be related to neurological immaturity &/or pharyngeal obstructions.
S/S: The increased frequency of apneic episodes is directly related to the degree of prematurity. Apneic episodes may result in bradycardia, hypoxia, respiratory acidosis, &/or death if not corrected.
Tx: SpO_2 monitoring, stimulation, manage the cause, theophylline or caffeine, doxapram, transfusion, CPAP, &/or mechanical ventilation.

Asthma   (See Chapter 2 Pulmonary Diseases)

Bronchiectasis   (See Chapter 2 Pulmonary Diseases)

Bronchiolitis
Inflammation of the bronchioles with airway obstruction. Usually caused by RSV infection, but some cases caused by adenovirus or parainfluenza virus. Often follows on upper airway viral infection. Additionally, it may be associated with asthma, tissue granulation or fibrotic tissue. Usually affects those under 2 years old, & primarily those under 6 months old.
S/S: Inspiratory wheezing & increased airway resistance which may progress to apnea. Chest congestion & mucus plugs.
Tests: CXR shows hyperinflation & areas of consolidation. FRC is increased. Clinical testing using Immunofluorescent Assay.
Tx: Humidification & oral decongestants for mild cases. More serious cases require bronchodilators, oxygen NC or croup tent, antibiotics (control secondary bacterial infections) and Ribavirin.

Bronchopulmonary Dysplasia (BPD)
An iatrogenic chronic lung disease that can develop in premature infants after a period of intensive respiratory therapy where positive pressure ventilation (high

144

peak pressures) and high oxygen are utilized. Barotrauma occurs with progressive vascular leakage, atelectasis, inflammation, & diffuse alveolar damage.

S/S: Significant respiratory distress. Pt usually suffering another pulmonary illness like HMD that required the mechanical ventilation & high oxygen. Right heart failure can also develop.

Tests: ABG reveal hypoxemia & hypercapnia secondary to airway obstructions &/or air trapping, & CXR may present with pulmonary fibrosis or atelectasis.

Tx: Prevention is best management tool by utilizing the lowest FiO$_2$ and the lowest amount of positive pressure for ventilation. Therapy is short-term support, oxygen, CPT, bronchodilators & steroids. Long term mechanical ventilation may be required.

Choanal Atresia

A congenital condition, with narrowing or blockage of the nasal airway by membranous or bony tissue. The cause of Choanal Atresia is unknown, but is thought to result from persistence of the membrane between the nasal (one side or both sides) & oral spaces during fetal development. Being obligate nose breathers, this condition can be viewed as an acute breathing problem.

S/S: A physical examination may show an obstruction of the nose, suction catheter may not pass through one or both nares, & neonate can not feed & breathe simultaneously.

Tests: An X-ray with dye or CT Scan with dye may show the occlusions.

Tx: Infants with bilateral Choanal Atresia may need resuscitation at delivery. An airway may be required to support breathing if infant can not tolerate mouth breathing. Rarely, intubation or tracheotomy may be required. Surgery to correct the disorder if required.

Congenital Diaphragmatic Hernia

The protrusion of the abdominal contents through the diaphragm, with abnormal placement affecting lungs & organs. Lung hypoplasia, decreased alveolar count, decreased pulmonary vasculature, pulmonary hypotension & unusual anatomy of the inferior vena cava, Bochdalek's Hernia (left lateral & posterior), Morgagni's Hernia (medial & anterior).

S/S, Tests: Severe respiratory distress, severe cyanosis, decreased breath sounds, scaphoid abdomen. Also, CXR will show displaced organs.

Tx: Establish & protect patient's airway with an ET tube, paralysis, HFV, ECMO & surgery to reposition organs.

Croup, Laryngotracheobronchitis (LTB)

A common acute upper airway viral disorder resulting in subglottic swelling & obstruction. Most commonly caused by the Parainfluenza virus, RSV or influenza virus. Croup cases worsened by bacterial superinfection with staphylococcus aureus, group A streptococcus pyogenes, or Haemophilus influenza. Usually affects those between 6 months & 3 years of age.

S/S: Symptomatic after 2-3 days of nasal congestion, fever & cough. Inspiratory stridor & significant barking cough. Dyspnea, cyanosis, exhaustion, & agitation occur.

Tests: ABG will reveal patient is hypoxic w/ increased levels of CO$_2$. X-ray to rule out Epiglottitis & confirm subglottic narrowing of trachea.

Tx: Depending on severity (i.e. breath sounds, cough, suprasternal retractions & cyanosis) home to hospital care is indicated. Care starting with cool mist, racemic epinephrine, steroids, or budesonide. Worsening cases may require mechanical ventilation.

Cystic Fibrosis (See Chapter 2 Pulmonary Diseases)

Epiglottitis

Acute inflammation of the epiglottis (supraglottic area). Can be life-threatening if complete upper airway obstruction develops. Most often caused by H. influenza Type B bacterial infection. Most common in those 2 to 8 years old.

S/S: Sudden onset fever, sore throat, cough, stridor, drooling (or difficulty swallowing), muffled voice, labored breathing, cyanosis. Can lead to coma & death if untreated.

Tests: Lateral neck X-ray shows enlarged epiglottis, with normal subglottic area. Auscultation will reveal diminished breath sounds & stridor.

Tx: Establish an airway (intubation/tracheostomy). Take care to intubate on first attempt & be prepared for emergency tracheostomy. The stimulation of intubation can cause complete obstruction. Sedation, antibiotic, oxygen, antipyretic. Extubation should be delayed until inflammation is resolved.

Heart Defects (Congenital)
Acyanotic Heart Lesions
Patent Ductus Arteriosus
Ventricular Septal Defect (VSD)
Atrial Septal Defect (ASD)
Coarctation of the Aorta
Cyanotic Heart Lesions
Tetralogy of Fallot
Transposition of the Great Vessels

Congenital Abnormalities. Presents in two categories, Acyanotic & Cyanotic.

S/S: Acyanotic heart disease presents as a left to right shunt.
         Cyanotic heart disease presents as right to left shunt.

Tests: Defect specific. X-rays.

Tx: Defect specific. Supportive. Possibly surgery.

Hyaline Membrane Disease (HMD), also known as
Infant Respiratory Distress Syndrome (IRDS) or
Respiratory Distress Syndrome (RDS)

A disorder of prematurity, manifested by respiratory distress due to decreased lung surfactant production. HMD/RDS almost always occurs in infants born before 37 weeks & the most premature have the greatest risk of the disease. Maternal diabetes, C-section & low birth weight are also predisposing factors.

S/S: Tachypnea, retractions, expiratory grunting, nasal flaring, cyanosis, & hypothermia. Pt usually deteriorates rapidly in the first 1 to 3 days of life, then should respond to appropriate therapy with marked improvement.

Tests: ABG will indicate hypoxemia & respiratory acidosis. CXR shows diffuse atelectasis. In Stage I RDS, reticulogranular (ground glass) pattern. Stage II RDS, reticulogranular (ground glass) pattern with air bronchograms centrally. Stage III RDS, the air bronchograms are diffuse. Stage IV RDS, diffuse atelectasis with lungs appearing white bilaterally.

Tx: Treat the atelectasis with CPAP or PEEP. Oxygen therapy, and mechanical ventilation if necessary. Also, sodium bicarbonate & glucose. HMD can be self-limiting, and usually lasts about 1 week. In more severe cases, HMD can lead to BPD.

Meconium Aspiration Syndrome (MAS)

Aspiration of amniotic fluid containing meconium, fetal lung fluid, transudate, and fetal urine.

S/S: Fetal tachycardia & absent fetal cardiac accelerations during labor. At birth low umbilical artery pH, APGAR <5 and meconium aspirated from trachea. Observation for MAS is warranted. Gasping respiration's, tachypnea, grunting & retractions.

Tests: CXR shows irregular pulmonary densities, atelectasis, hyperlucent areas (air trapping). ABG reveals hypoxemia with mixed acidosis.

Tx: Suction oropharynx at delivery, deliver humidified oxygen, insert ET tube for gentle suction. Remove ET tube & inspect for meconium. Repeat if tube is stained with Meconium (2 to 4 times). If patient's condition worsens utilize CPAP, or intubate & start mechanical ventilation if indicated.

Persistent Fetal Circulation or
Persistent Pulmonary Hypertension of the Newborn (PPHN)

A complex syndrome with characteristics including: pulmonary hypertension & hypoxemia, secondary to right to left shunt through foramen ovale and ductus arteriosus. PVR is increased from pre- or post-natal asphyxia.

S/S: Hypoxemia, cyanosis, tachypnea, acidosis.

Tests: CXR & $PaCO_2$ measurements. Infants with significant shunt through the ductus arteriosus can be found by comparing two SpO_2 monitors, one on the right arm & the other on either leg.

Tx: Supportive care using oxygen (for hypoxemia), surfactant (for IRDS), glucose (for hypoglycemia), & inotropic agents (for cardiac output or systemic hypotension). If not corrected, pt may be indicated for Nitric Oxide & HFV via mechanical ventilator (with sedation/paralysis to reduce pain & anxiety, which contributes to PPHN). Persistent cases man require the use of ECMO.

Respiratory Distress Syndrome (RDS or IRDS)
see Hyaline Membrane Disease

Respiratory Syncytial Virus (RSV)

RSV is one of the most significant causes of lower respiratory illness (bronchiolitis & pneumonia) & can be fatal in infants & young children. Premature babies, patients with immune system problems, or heart/ lung problems may be at high risk. RSV is a highly contagious annual disease, which occurs late in autumn or winter. It spreads via secretions (saliva or mucus) when they cough, or sneeze. Normal healthy children & adults present with milder symptoms & patients cannot develop full immunity to the virus.

S/S: Dyspnea, cough & wheezing are the most prominent symptoms. RSV affects the nose, throat & upper respiratory system often presenting like pneumonia, laryngotracheobronchitis (LTB) or bronchiolitis. Symptoms include mild sore throat, cough, stuffy or runny nose, earache, & fever.

Tests: CXR may indicate bronchopneumonia if present. Laboratory test show a normal leukocyte count but granulocytes may be elevated.

Tx: RSV usually resolving itself in 10-14 days. Ribavirin (antiviral) can accelerate recovery. Depending on severity & complications, respiratory support may be required. Secondary bacterial infections are treated with antibiotics & fluids to prevent dehydration.

Sudden Infant Death Syndrome (SIDS)
SIDS is the leading cause of death of patients 2 weeks to 1 year old & remains unexplained. Unexpected death while sleeping thought to be due to dysfunction of neural cardiorespiratory control mechanisms.

S/S: No symptoms have been identified, although risk can be assessed through patient & parent histories.

Tx: Infant & apnea monitoring along with family training. Don't let infant sleep on stomach while unattended.

Tracheoesophageal Fistula (T-E Fistula)
A fistula from near the carina to the lower esophagus. 5 types of fistulas exist: Atresia (esophagus separation), Fistula only (trachea & esophagus connected to both stomach & lungs), Atresia plus upper fistula (trachea & esophagus connected above carina), Atresia plus lower fistula (trachea & esophagus connected below carina), & Atresia plus double fistula (trachea & esophagus connected above & below carina).

S/S: Excessive secretions, coughing, & cyanosis. Depending on fistula type air can be forced into the stomach &/or aspiration pneumonia reflux.

Tests: CXR, CT and/or MRI Scans are used to identify & diagnose.

Tx: Oxygen, ventilation support & surgery to correct the connections.

Transient Tachypnea of the Newborn (TTNB)
Respiratory distress with transient tachypnea & hypoxemia caused by delayed absorption of fetal lung fluid. Predisposing factors are C-section (lack of vaginal squeeze) and prematurity. Increased airway resistance & decreased lung compliance resulting in hyperinflation &/or air trapping.

S/S: In the first few hours respiratory rate increases (up to 120). Respiratory distress, grunting, nasal flaring, retractions & may be cyanotic.

Tests: CXR reveals lungs with infiltrates & heavy perihilar markings, fluid is often present in fissures, & lung periphery remains clear.

Tx: Usually self-limiting with recovery in 2 to 5 days. Pt may need oxygen therapy via oxyhood, or NC, or CPAP while monitoring via ABG's or transcutaneous monitoring. Mechanical ventilation is rarely required.

Chapter 18

BLS & ACLS

BLS

Check for responsiveness first, then open the airway, check for breathing, &
check for circulation (pulse).
Shake victim gently & shout, "Are you OK?"
Ask, "Are you choking?"
If victim cannot speak or breathe: Activate EMS. Call 911. If alone, yell for help!
Open the airway.

Check for pulse. Also check pulse every few minutes during CPR. If an adult is
found unresponsive, call 911 immediately. If a child or infant is found
unresponsive, and you did not witness the event, and you are alone, perform
resuscitation for one minute, and then call 911.

Follow chart below for ABCD maneuvers.

BLS ABCD Maneuvers for Adults, Children, and Infants

MANEUVER	ADULT over 8 years old	CHILD age 1 to 8	INFANT* under 1 year of age
A- Airway	Head tilt-chin lift (if suspected trauma or spinal injury, use jaw thrust)		
B- Breathing Look, listen, & feel. Watch for chest rise.(Reposition & retry if necessary)	2 breaths @ 1 second/breath	2 effective breaths @ 1 second/breath	
Rescue breathing without chest compressions (for those w/ a pulse)	10 – 12 breaths/ minute (appx.)	12 – 20 breaths/ minute (appx.)	
Rescue breaths for CPR w/ advanced airway	8 – 10 breaths/ minute (appx.)		
FBO - Foreign Body Airway Obstruction (If FBO present, repeat breaths, sweeps, & thrusts until successful or unconscious. If	Only perform finger sweep if FBO is visible. Perform abdominal thrusts. (if obese or pregnant, perform chest thrusts instead)		5 Back slaps & 5 chest thrusts

unconscious, perform CPR.)			
C- Circulation Pulse check (10 seconds or less)	Carotid		Brachial or femoral
Compression landmarks	Lower half of sternum, between nipples		Just below nipple line (lower half of sternum)
Compression method Push hard & fast Allow complete recoil	Heel of one hand, other hand on top	Heel of one hand, or as for adults	2 or 3 fingers (or if 2 rescuers, can use 2 thumb-encircling hands)
Compression depth	1.5 to 2 inches		Appx. 1/3 to 1/2 the depth of the chest (this is appx. 1 to 1.5 inches for a child, and 0.5 inch to 1 inch for an infant)
Compression rate	Appx. 100 / minute		
Compression-ventilation ratio	30:2 (one or two rescuers)	30:2 (single rescuer) 15:2 (two rescuers)	
D- Defibrillation (AED) Automated External Defibrillator	Use adult pads (do not use child pads on adults)	Use AED after 5 cycles of CPR (out of hospital). Use pediatric system for child 1-8 years if available. For sudden collapse (out of hospital) or in-hospital arrest, use AED as soon as available.	No recommendation for infants under 1 year of age

*Newborn information not included.

Accessed AHA BLS charts: February, 2007; includes updates for 2005.

ACLS Pulseless Arrest

1. BLS Algorithm: Call for help; give CPR.
   Give Oxygen when available.
   Attach Monitor/defibrillator when available.

2. Check rhythm: Is it a shockable rhythm?

3. Yes- shockable. VF/VT. Go to step 4.
   No- not shockable. Asystole/PEA. Go to step 9.

4. Give 1 shock: Manual biphasic; device specific (usually 120 to 200J).
   　　　　　　　　AED: device specific.
   　　　　　　　　Monophasic: 360J.
   　　　　　　　　Resume CPR immediately after the shock, with 5 cycles of CPR.

5. Check rhythm: Is it a shockable rhythm?
   　　　Yes- Go to step 6.
   　　　No- Go to step 12.

6. Continue CPR while defibrillator is charging.
   Give 1 shock:
   　　　Manual biphasic; device specific, (usually 120 to 200J) (if unknown, give 200J).
   　　　AED: device specific
   　　　Monophasic: 360J.
   　Resume CPR immediately after the shock.
   　When IV/IO available, give vasopressor during CPR (pre or post shock).
   　　　Epinephrine 1 mg IV/IO & repeat every 3 to 5 minutes
   　　　or May give 1 dose of Vasopressin 40 U IV/IO to replace 1st or 2nd dose of Epi.

7. Yes- shockable. VF/VT. Go to step 8.
   No- not shockable. Asystole/PEA. Go to step 11.

8. Continue CPR while defibrillator is charging.
   Give 1 shock:
   　　　Manual biphasic; device specific, (usually 120 to 200J) (if unknown, give 200J).
   　　　AED: device specific
   　　　Monophasic: 360J.
   Resume CPR immediately after the shock.
   Consider antiarrhythmics; give during CPR (pre or post shock).
   　　　Amiodarone (300 mg IV/IO once, then consider additional 150 mg IV/IO once;
   　　　　　　or
   　　　Lidocaine (1 - 1.5 mg/kg 1st dose, then 0.5 to 0.75 mg/kg IV/IO,
   　　　　　　　　max 3 doses or 3 mg/kg)
   Consider Magnesium, loading dose 1 – 2 g IV?IO for torsades de pointes.
   After 5 cycles of CPR, repeat #5, and continue from there.

....

9. Step #9 is for non-shockable rhythm (Asystole/PEA).
   Resume CPR for 5 cycles.
   When IV/IO available, give vasopressor.
   　　　Epinephrine 1 mg IV/IO & repeat every 3 to 5 minutes.
   　　　or May give 1 dose of Vasopressin 40 U IV/IO to replace 1st or 2nd dose of Epi.
   　Consider Atropine 1 mg IV/IO for asystole or slow PEA rate,
   　　　repeat every 3 to 5 minutes (up to 3 doses).
   　Give 5 cycles of CPR.

10. Check rhythm: Is it a shockable rhythm?
   　　　Yes- Go to step #4.
   　　　No- Go to step #11.

11. If asystole or PEA, go to step #9.
     If pulse present, begin post-resuscitation care.

*J indicates Joules.
*PEA indicates pulseless electrical activity.

For CPR cycles, see CPR chart in this chapter.
During CPR, HCP's should secure airway & confirm placement.
After the advanced airway is in place, rescuers do not deliver cycles of CPR.
Rescuers should give continuous chest compressions without pauses for breaths.
Give 8-10 breaths/minute.
Check rhythm every 2 minutes. Rotate compressions with the rhythm checks Q2 minutes.
Find & treat any potential contributing factors including:
  Hypothermia, hypoglycemia, hypokalemia, hyperkalemia, hypovolemia, hypoxia,
  hydrogen ion (acidosis). Tamponade (cardiac), thrombosis (coronary or pulmonary),
  tension pneumothorax, trauma, toxins.

Many algorithms exist for ACLS. These can be accessed online for the most updated
charts. This chart is a summary of Pulseless Arrest. Accessed data online February,
2007.

NOTES:

Chapter 19

EQUIPMENT DISINFECTION

EQUIPMENT DISINFECTION METHODS:

CLEANING

- Removes all soil, particles, and foreign material.
- Equipment is best cleaned by pre-soaking, then washing in mild fragrance-free detergent, using brushes or lint-free towels.
- Equipment must be disassembled & cleaned prior to disinfection or sterilization. (Equipment may also need to be dried, depending on method of disinfection/sterilization to follow.)

DISINFECTION

- Process used to destroy most pathogenic microorganisms, except spores. However, spores can be destroyed by certain high level disinfectants with adequate exposure time, making some disinfectants true sterilization agents.
- Disinfection prevents infection and cross contamination, and it is a good process for most respiratory equipment. High level disinfection or sterilization is needed for invasive equipment.
- Equipment is usually rinsed & air-dried after disinfection (depending on disinfection method used).

Low Level & Medium Level Disinfectants: (non-sporacidal)
Alcohols
Iodophors (Iodine mix; also an antiseptic)
Pasteurization (does remove some spores)
Phenolics
Quaternary Ammoniums
Acetic Acid - Vinegar Good disinfectant for home healthcare use. Articles are washed, rinsed, then soaked for 30-60 minutes in vinegar diluted with water to make a 1:3 solution. After disinfection, articles are rinsed again, then air dried.

Common High Level Disinfectants: (some are sporacidal as noted)

Gluteraldehyde 2% solutions. Trade names include Cidex®, Sonacide®, Sporicidin®, Hospex®, Omnicide®, Metricide®, and Wavicide®. Cidex usage for Disinfection: 10 minutes. Sterile/Sporacidal: 3 hours. Sonacide usage for Disinfection: 10 minutes at room temp, 5 minutes at 60°F. Sterile/Sporacidal: 60 minutes at 60°F. (Sonacide can be used in an ultrasonic cleaner.)

Bleach (Sodium Hypochlorite) Disinfection: dilute with water 1:10 for quick surface cleaning, blood spills, or home use. Sterile/Sporacidal: soak in 1:50 solution (about 1000 ppm) for 10 minutes.

Hydrogen Peroxide 6% solution. (also an antiseptic). Disinfection: 10 minutes. Sterile/Sporacidal: 6 hours at 20ºC.

Peracetic Acid 0.35%. Specially formulated anticorrosive solution is a disinfectant and is sporacidal.

Ultrasonic Cleaner This can be utilized as both a low & high level disinfectant; sporacidal with proper sterilization agent.

**Note: Exposure time (soaking in solution) to reach disinfection and sterilization varies by product, and is subject to change.

STERILIZATION

- Process used to completely destroy &/or remove all forms of microbial life, including spores.

Sterilization Methods include:

High Level Disinfectants (as listed above)

Autoclaving (utilizes steam @ 120°C under pressure for 15-45 min)

Boiling (more often used in homes; 212°F for 15 min or longer)

Dry Heat (utilizes temps of 170-180°C for 2-3 hours)

Ethylene Oxide (important to clean & completely dry equipment first; temps & exposure times vary greatly)

Incineration (for disposable articles only)

Ionizing Radiation (Gamma rays; high energy wave radiation)

**Note: Always refer to manufacturer's instructions & employer's policy regarding proper usage of products for each type of equipment. Also, follow guidelines on dilution, duration of use, testing of solution, cautions, & hazardous material information.

Chapter 20

Misc. FORMULAS & CONVERSIONS

Gas Laws

Avogadro's hypothesis: Equal volumes of different gases at equal temperature contain the same number of molecules. Equal numbers of molecules in the same volumes at the same temperature will exert the same pressure. (One mole of any gas will contain 6.02×10^{23} molecules and will occupy a volume of 22.4 L at a temperature of 0°C and a pressure of 760 mmHg.)

Boyle's Law: $P_1 V_1 = P_2 V_2$ (at constant temperature)

Charles' Law: (at constant pressure) $V_1/T_1 = V_2/T_2$

Combined Gas Law: (at a constant mass) $P_1 V_1 / T_1 = P_2 V_2/T_2$
Dalton's law: In a gas mixture the pressure exerted by each individual gas in a space is independent of the pressures of other gases in the mixture. e.g. $P_{Alv} = P_{H2O} + P_{O2} + P_{CO2} + P_{N2}$
$\quad P_{Gas1} = \%$ of total gases x P_{Tot}

Gay-Lussac's Law: (at constant volume) $P_1/T_1 = P_2/T_2$

Graham's Law: The rate of diffusion of a gas into a liquid is inversely proportional to the square root of its molecular weight.

Henry's Law: The amount of a gas dissolved in a liquid is directly proportional to the partial pressure of the gas above the liquid.

Ideal gas Law: $PV = nRt$; Kinetic theory of gas behavior.

Temperature Conversion:

Convert Fahrenheit (°F) to Celsius (°C)
°F = (°C x 1.8) + 32

Convert Celsius (°C) to Fahrenheit (°F)
°C = (°F − 32) / 1.8

Celsius	Fahrenheit
0	32
35	95
36	97

37	98.6
38	100.4
39	102
40	104

Torr, cmH$_2$0, mmHg, & kPa Conversions:

1 mm Hg = 1 torr = 1.36 cm H$_2$0

torr x 1.36 = cm H$_2$0
cm H$_2$0 / 1.36 = torr

1 kPa = 7.5 torr = 10.2 cm H$_2$0
(kilopascal or kPa is the SI standard unit of pressure)

Metric Conversions:

Microgram (mcg or µg)	100 mcg = 0.1 mg 1000 mcg = 1 mg
Milligram (mg)	1 mg = 0.001 gram 100 mg = 0.1 gram = 0.0001 kg = .0035 oz 1000 mg = 1 gram = 0.001 kg = 0.035 oz
Milliliter (mL)	1 mL = 1 cc = 20 drops 30 mL = 1 oz = 28 grams 1000 mL = 1 Liter
mL household measures	1 tsp = 5 mL 1 Tbsp = 15 mL 1 cup = 240 mL = 0.24 Liters = 8 oz 1 pint = 474 mL = 0.47 Liters = 16 oz 1 quart = 946 mL = 0.95 Liters = 32 oz 1 gallon = 3.79 Liters = 128 oz

Kilograms (kg) & Pounds (lb)	Pounds / 2.2 = Kilograms Kilograms x 2.2 = Pounds
	1 lb = 0.45 kg = 454 grams = 16 oz 1 kg = 2.2 lb = 1000 grams = 35.3 oz

NOTE: Numbers are rounded where appropriate to improve readability & speed calculation time.

HEIGHT & WEIGHT CONVERSIONS

Height Inches (in) to Centimeters (cm) Conversion

Convert Inches to Centimeters: **in x 2.54 = cm**
Convert Centimeters to Inches: **cm / 2.54 = in**

Inches	cm		Inches	cm		Inches	cm
30	76		60	152		70	178
40	102		62	157		72	183
50	127		64	163		74	188
54	137		66	168		76	193
58	147		68	173			

IBW (Ideal Body Weight)

Female IBW = 105 + (5 x height in inches over 60)
Male IBW =   106 + (6 x height in inches over 60)

Examples: 63 inch tall female has an IBW of 120 lb: 105 + (5 x 3)
            70 inch tall male has an IBW of 166 lb: 106 + (6 x 10)

BSA (Body Surface Area m^2)

Calculation: ($\sqrt{}$ square root symbol)

   $\sqrt{}$ **(Height (in) x Weight (lb)) / 3131**
or
   $\sqrt{}$ **(Height (cm) x Weight (kg)) / 3600**

Example: The BSA of a 66 inch tall 135 lb female is: **1.7 m^2**

BMI (Body Mass Index)

BMI = (Weight in lb x 700) / (height in inches^2)

Sample: Find the BMI of a 170 lb male who is 70 inches tall
  (170 x 700) / (70 x 70)
  (119,000) / (4,900) = 24.2

BMI Values are related to Nutritional Status as follows:
  BMI < 20   Underweight
  BMI 20-25 Normal weight
  BMI 25-30 Overweight
  BMI > 30   Obese

BMR (Basal Metabolic Rate)
& REE (Resting Energy Expenditure)

BMR is the measure of ones energy requirements in calories per hour. REE estimates the daily caloric requirements. This is usually obtained after 10 hours of fasting. This can be obtained by indirect calorimetry that measures VO_2 & VCO_2, or by using formulas. Listed here is one of the quicker calculations to estimate daily caloric requirements:

IBW (in pounds) x activity factor x illness factor

Multiply desired weight or IBW (in pounds) by activity factor as follows:
Activity factors:
  12 = Sedentary (sitting or laying down most of the day; very little walking; no regular exercise routine; most patients are within the sedentary range).
  15 = Moderately active (plenty of daily walking, plus moderate exercise for ½ hour or more, at least 3 times per week).
  18 = Vigorously active (lots of daily activity & daily strenuous exercise routine, like jogging, skiing, fast biking, or fast swimming).

Next, multiply the result by the illness factor as follows:
Illness factors:
  1.0   for healthy person with no current illness
  1.15 for mild illness
  1.3   for moderate illness
  1.5   for severe illness, pregnancy, or lactation.

INDEX

REFERENCES:

1. Wilkins, Robert L., Scanlan, Craig L., Stoller, James K.: Egan's Fundamentals of Respiratory Care, 8th edition. 2003. Elsevier Science.
2. Gold, Warren M., Nadel, Jay A.: Atlas of Procedures in Respiratory Medicine. 2002. WB Saunders Company.
3. Wyka, Kenneth A., Mathews, Paul, Clark, William F., et al.: Foundations of Respiratory Care. 2001. Delmar Learning.
4. U.S. Center for Disease Control & Prevention (CDC). On-Line: www.cdc.gov. Accessed March 2006.
5. Czervinske, Barnhart, Sherry.: Perinatal and Pediatric Respiratory Care. 2002. WB Saunders Company.
6. Murray, John F., Nadel, Jay A.: Textbook of Respiratory Medicine, 3rd edition. 2001. WB Saunders Company.
7. PDR Staff Physicians: Physicians' Desk Reference, 58th edition. 2004. Medical Economics Company.
8. Damico, Christine M., RN, MSN, CPNP, et al.: Nursing 2004 Drug Handbook. Springhouse Corporation.
9. Giddens, Jean Foret, PhD, MSN, RN, CS, Langford, Rae W., EdD, RN: Nursing PDQ. 2004. St Louis, Mosby.
10. Myers, Ehren, Hopkins, Tracey: LPN Notes. 2004. Philadelphia, F.A. Davis.
11. Lewis, Sharon, Dirksen, Shannon Ruff, Heitkemper, Margaret. Medical Surgical Nursing. 2003. Elsevier Science.
12. Martin, Kevin, RRT. Neonatal Respiratory Care: Crisis Management. 2003. RCECS.
13. Corning, Helen Schaar, RRT, Bryant, Stanley L. CRT: Mosby's Respiratory Care PDQ. 2006. Elsevier Science.
14. LaSage, Paul EMT-P, Assistant Chief, Derr, Paula, RN, BSN, CCRN, CEN, Tardiff, Jon, Paramedic, EMS Field Guide ALS Version, 13th Edition 2001. InforMed.
15. Atlas of Pathophysiology, Judith Schilling MaCann, RN, MSN, Lippincott, 2nd edition, 2006
16. LaSage, Paul EMT-P, Assistant Chief, Derr, Paula, RN, BSN, CCRN, CEN, Tardiff, Jon, Paramedic, EMS Field Guide ALS Version, 13th Edition 2001. InforMed.
17. WebMD Health (on-line) www.webmd.com
18. American Lung Association 1-800-330-5864
19. American Cancer Society 1-800-ACS-2345
20. Center for Disease Control (CDC) (on-line) www.cdc.gov

Notes on this collaborative effort book:
Chapters written by Helen Schaar Corning RRT include:
Chapters 2, 3, 4, 5, 6, 7, 8, 9, 10, 13, 14,15, 16, 18, 19, and 20.

Chapters written by Stanley L. Bryant Jr., include:
Chapters 1, 11, 12, 17 and all the graphics in this book.

Data-Stat Respiratory Therapy Desk Reference

Chapter Index:

www.ingramcontent.com/pod-product-compliance
Lightning Source LLC
Chambersburg PA
CBHW032020170526
45157CB00002B/785